In Focus

Photographic Evidence of
Unconventional Flying Objects

MAX –

THE TRUTH IS OUT THERE

N. Roesler

PARACon 2010

NICHOLAS D. ROESLER

Cover design by Garret Moore
Page design by Karen McChrystal

ISBN: 978-0-615-23482-3

RAVENWOOD
PRESS

Brookfield, WI

Printed in the United States of America

FOR TRACEY

ACKNOWLEDGMENTS

This book is the product of independent research conducted by countless individuals over a number of years, but I would like to thank, specifically, the following researchers and organizations, who have contributed information, interviews, photographs, case reports, anecdotes, and their professional opinions.

Before all else, there are a few people whose involvement warrant special attention.

First and foremost, I am indebted to the women in my life, without whom none of this would have been possible. My life is infinitely better thanks to your love and light. Before anyone else, this book is for them – my mother, Linda Roesler, my sister Erin, and the woman who will not only be my wife, but forever my best friend, Tracey Ann Nicely. I love you all.

Next, my eternal gratitude goes out to the man who has been my friend, my colleague, and my surrogate brother for so many years, Jon Nowinski. You are, perhaps, the living embodiment of dogged determination, steadfastness, and steeled professionalism that we all strive for in this field. I would not be who I am today were it not for your presence. I am truly honored to count you among my inner circle.

Thirdly, my deepest thanks go out to fellow researcher and author Ruben Jose Uriarte. Your friendship and respect have been the true impetus behind this project, and I will be forever grateful for all you've done.

My sincerest thanks to Dr. Jack Kasher, for his consideration in penning the introduction to this book. He is an excellent scientist and researcher, and it means a great deal to me that he thought highly enough of my work to support it by adding his endorsement.

Lastly, I want to thank in the strongest terms possible, my colleague, editor, and dear friend Pamela Ward. Without your friendship, faith in my abilities, encouragement along the way, and assistance throughout the process, this project wouldn't be what it is today.

Of all the people I've met in my years of research and the lifelong friends I've made, I am particularly privileged to call you not only my colleague, but my dear, dear friend. Thank you so much for all you've done. I will never be able to thank you enough for all you've done.

Additionally, I would like to thank, in alphabetical order:

Dino Brancato, Dr. Donald R. Burleson, James Carrion, John Robert Colombo, Scott Cossel, Richard and Karyn Dolan, Ann Druffel, Stanton Friedman, Christian Goos, John Greenewald, Jr., Bill Konkolesky, Dr. Bruce Maccabee, Garret Moore, Harold and Debbie Nicely, William, Rebekah, Aubrey, and Gavin Nicely, Alejandro Rojas, Chris Rutkowski, John and Kathy

Schuessler, Robert and Lulin Simpson, Cathy Spitzenberger, Robert and Susan Swiatek, David Twichell, Charles Ward, Bruce A. Widaman, and Dr. Robert Wood.

Thanks are also due the following organizations, for opening their archives and sharing their expertise:

The Black Vault, The Center for UFO Studies, The Fund for UFO Research, The Mutual UFO Network, The Smoking Gun Research Agency, the University of Arizona at Tucson Library – Special Collections Section, the University of Texas at Arlington Library – Special Collections Section, and the Gerald R. Ford Presidential Library.

This book would not be possible without the assistance and opinions of the experts in this field in concert with six decades' worth of technical, scientific, and anecdotal data.

It has been said that extraordinary claims require extraordinary evidence as to their validity. I believe that the scientific and technical data contained herein, along with the testimony of so many researchers, all at the pinnacle of their respective fields, meets that burden of proof.

Nicholas Roesler

Milwaukee, Wisconsin

IN FOCUS

PHOTOGRAPHIC EVIDENCE OF UNCONVENTIONAL FLYING OBJECTS

CONTENTS

Chapter 5. The Roswell Incident

Chapter 6. The Trent Photographs

Chapter 7. The Washington National Airport Sightings

Chapter 11. The Yorba Linda, California Photograph

Chapter 12. The Colfax, Wisconsin Daylight Disk Photos

Chapter 13. Vancouver Island, British Columbia, Canada

Chapter 19. In Summation   207

About the Author   217

Introduction

As the old proverb says, *"A picture is worth a thousand words."*

Properly documented photographs and videotapes can provide strong evidence for a wide range of phenomena. Video cameras on police cars and in and around stores, as well as videos taken by private citizens, have been invaluable in proving or disproving allegations where the commission of a crime is suspected. Individuals might walk around their houses with a video camera, documenting their possessions in case of a fire. Any number of other examples of the usefulness and importance of photographs and videos could be listed.

Here we are concerned with their use in a specific area – namely in the investigation of UFO sightings. Visual evidence for UFOs goes back many, many years, even to drawings inside the pyramids. But an actual photo or video of a UFO, when properly documented, is obviously much more compelling than a drawing when an anomalous event has been observed. Photographic evidence is an extremely important part of the investigation of UFOs, and strengthens a case considerably when it is present and well documented. So a welcome addition to the UFO literature would be a thorough examination of the basic principles of photography, and how they should be properly used in the investigation of UFO cases. In other words, we need a user's manual, so to speak, for the correct methodology in applying photography to the study of UFOs.

With this excellent book, Nick Roesler has provided us with our user's manual. He is eminently equipped to discuss and substantiate the importance and necessity of photography in analyzing UFO cases. He is the Staff Photographer for the *MUFON UFO Journal*, an international journal dedicated to the scientific study of UFOs. In addition, he is a UFO investigator and photographic consultant for the Smoking Gun Research Agency, and has been a member of the National Press Photographers' Association. He has also done freelance photography work on a variety of subjects.

Nick begins with a brief history of photography in chapter one, and follows with the basic ground rules as they apply to UFO investigations in chapters two and three. Then in the next fifteen chapters he shows how these principles have been used in fifteen separate cases, dating from the Battle of Los Angeles in 1942 up to the O'Hare disc sighting in November, 2006. These cases cover a wide range of sightings, and he examines several different types of photographic evidence. In the Washington National Airport case (Chapter 7), the UFO was caught on radar. The Heflin and Colfax Wisconsin photos (Chapters 8 and 12) were Polaroid shots. There were regular photos from the Battle of Los Angeles (Chapter 4), Trent (Chapter 6), Michigan (Chapter 9), Yorba Linda (Chapter 11), Vancouver Island (Chapter 13), Gulf Breeze (Chapter 14), the Phoenix Lights (Chapter 16), Millstadt, Illinois (Chapter 17), and O'Hare International (Chapter 18) cases. The 1991 Shuttle case (Chapter 15) was an analysis of a videotape. Finally, the Roswell incident (Chapter 5) dealt with an analysis of the Ramey memo taken from a photograph. Along the way, Nick cites the work of Dr. Bruce Maccabee and Dr. Richard Haines, two of the best in the UFO field at analyzing photographs, and gives credit to the Fund for UFO Research, which has funded many photo-analytic studies of UFO cases.

So Nick has done an excellent job in providing us with a badly needed reference for the use of photographic analysis in the study of UFO cases. In doing so he has revisited many of the famous cases from the past, and added his own personal, well-reasoned ideas to the mix. This book is well worth your time, both as a welcome reference and as an instructive tool for understanding the place of photography in the study of UFOs.

Dr. Jack Kasher
Omaha, Nebraska

1. The Origins of Modern Photography

To say the name Leonardo da Vinci is to immediately conjure the image of the master responsible for such classic works of art as the *Universal Man, Mona Lisa*, and the *Last Supper*. Leonardo, however, was much more than the man behind the paintings for which he is most often remembered. Leonardo was, in fact, predominantly a man of science, and quite possibly the greatest mind of the Renaissance.

Leonardo's self-portrait, circa 1512
Courtesy Wikimedia Foundation

In addition to the commissions he earned serving royalty and the church, Leonardo was a scientist, engineer, mathematician, painter, sculptor, architect, musician, poet, and botanist.[1] He is, in fact, the person for whom the term *"renaissance man"* seems to have been created. The term, reserved for those who have extraordinary talent in multiple arenas, seems uniquely suited to an individual such as Leonardo da Vinci. He was the type of unparalleled genius who seems to become the master of whatever discipline, simply due to intellect and will power.

While history remembers Leonardo chiefly as a painter and sculptor, his writings and inventive nature are perhaps his greatest legacy to the worlds of art and the sciences. The genius who gave the world some of its most famous works of art is also responsible for penning manuscripts and journals on subjects as diverse as anatomy, art, botany, architecture, veterinary medicine, and even warfare.[2] It is important to note that, in his time, Leonardo was best known as an engineer. It was, quite simply, his job to take a physical object, mentally strip that object down to its individual parts, and reconstitute it in a way which was, if only to Leonardo himself, simpler and more efficient.

Leonardo da Vinci and the Atlantic Codex

Leonardo wrote obsessively virtually all his life. Of the four thousand individual pages that would come to be known as his "notebooks" or "codices", many of them contained sketches and schematic drawings for items that seemed fantastic in the fifteenth and sixteenth centuries, but are seen as commonplace today.

It is the matter of Leonardo's notebooks that bear out his standing in the scientific community, even prior to the formal recognition of the scientific method.

Writing of Leonardo's impact on empirical science, noted physicist and Leonardo researcher Dr. Fritjof Capra, in his 2007 work, *The Science of Leonardo*, states:

"In the intellectual history of Europe, Galileo Galilei, who was born one hundred twelve years after Leonardo, is usually credited with being the first to develop this kind of rigorous empirical approach and is often hailed as the 'father of modern science'. There can be no doubt that this honor would have been bestowed on Leonardo da Vinci had he published his scientific writings during his lifetime, or had his Notebooks been widely studied soon after his death." [3]

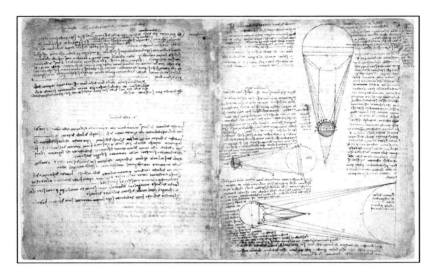

Leonardo's Codex Leicester, circa 1506
Courtesy: Seth Joel/CORBIS

Dr. Fritjof Capra states, again in *The Science of Leonardo*, that simply by reading through Leonardo's notebooks, it is evident that Leonardo was of superior intellect, even compared to his contemporaries. Discussing the matter of engineering, Dr. Capra states:

"Drawings of similar machines were produced by other Renaissance engineers. However, as art historian Daniel Arasse points out, while theirs were merely explanatory, Leonardo's are convincing, persuading the viewer of the feasibility and soundness of the author's designs." [4]

Centuries ahead of his time, Leonardo was kept from realizing his vision simply due to the lack of the requisite materials being readily available to complete many of the inventions found within the pages of the Codex Atlanticus, Codex Leicester, and others. Items Leonardo envisioned, but never lived to realize, included various types of aircraft, including the modern helicopter and hang glider, as well as more technical equipment. One such example can be found in Leonardo's Codex Atlanticus, a twelve volume collection of writings spanning in excess of 1100 pages, which he supposedly began compiling from 1478 until his death in 1519.[5]

The Atlantic Codex and the camera obscura

The Codex Atlanticus holds an updated schematic and design for the camera obscura, the simple pinhole camera originally designed some four hundred fifty years earlier by Muslim scientist Ibn al-Haytham[6], whose early experiments with regard to optics were set forth in his aptly titled *Book of Optics*. It is in his writings that the camera obscura literally receives its name, in Arabic, *Al-bayt al Muthlim*, or "dark room" in English.

Ibn al-Haytham

Although the principles behind the pinhole camera have been known since ancient times, it is the research of Leonardo that truly gave birth to the principles that would lay the foundation for modern photography.

It was Leonardo who first adapted the design of the camera, based on the *Book of Optics*, later demonstrating how an image is formed within the eye, simply using the camera obscura as an analog. It is noteworthy that Leonardo's interpretation of the *Book of Optics*, which, in turn, led to the design of the camera obscura, as well as writings positing the chemical processes by which to create latent images, came nearly four hundred years before the commonly accepted advent of photographic technology elsewhere in Europe.

The camera obscura in principle

The camera obscura, or "pinhole camera" as originally conceived, was used primarily in sketching and drawing. This camera, lacking the lens of a conventional camera, can focus all of the visible light from the scene to a single point, just as Leonardo had predicted in the late fifteenth century.

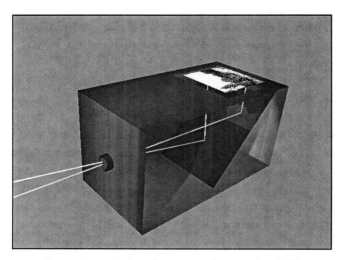

Computer rendering of a camera obscura, circa 1830
Courtesy Wikimedia Foundation via GNU Free Documentation License

Although it was Leonardo who expanded on the principle, devising a viable blueprint for the camera, it was not until the mid-seventeenth century, long after Leonardo's death, that the design for the simplest camera obscura was further improved upon, with the convex lens first replacing the simple pinhole sometime in the 1760s.

Prior to the advent of latent image photography, cameras were the tools of artists, primarily to aid in sketching, although variations on the standard design were used by the astronomers of the day to safely view celestial events, such as solar eclipses.

The modern era

The modern photographic process which we know today resulted from the combination of several technical and scientific breakthroughs over a period of several hundred years, including the discoveries of silver nitrate by Albertus Magnus and silver chloride by Georges Fabricus, the sensitivity of those chemicals to light, and the refining of various processes by which to fix a visible image.

The very word *"photography"* comes to the English language from French, with the French translation coming originally by way of the Greek words *phos* (light) and *graphê* (drawing).[7]

Photography as a practical process dates to the early 1820s with the development and refinement of these chemical processes. The first modern photographic image was produced in France during the summer months of 1827 by French inventor and scientist Nicéphore Niépce, in concert with Louis Jacques Daguerre who, until that time, was best known as a painter and architect.[8] This first photograph took over eight hours' exposure time to capture. Niépce and Daguerre expanded on earlier research, such as earlier experiments by Niépce, which used silver as catalyst.[9] This process would later be replicated and, in turn, expanded upon, by contemporaries of Niépce and Daguerre, including William

Fox Talbot, who developed the *calotype* process, which produces a negative image, and John Hershel, who pioneered the *cyanotype* photographic process in the 1840s, to compete with Daguerre's self-named *Daguerrotype* photo process, which was developed and patented by Daguerre in 1839, following his partner's death six years earlier in 1833.

Remnants of Talbot's process can be seen in modern photography to this day, as he went on to pioneer its use as an artistic medium. Hershel's process, the cyanotype, is better known today simply as the blueprint.

The first known fixed photograph, produced by Niepce and Daguerre, 1827.
Courtesy Wikimedia Foundation

Modern medium and large format photography

The medium and large photographic formats are formats that harken back to the earlier days of photography, particularly toward the dawn of the twentieth century.

The term *'medium format'* traditionally refers to any film size which falls between the more prevalent 35mm size and *'large format'*, for which the common negative size is 4x5 inches.

The larger format (in comparison to the 35mm format, with which most are familiar) has the advantage of higher image resolution, making the medium and large formats particularly attractive to artists and those who made their living as photojournalists.

While the larger formats allowed for heightened image resolution, the commercial availability of film, processing, and price per unit makes large format photography prohibitively expensive outside of the artistic community, where a number of art photographers work specifically with large format equipment.

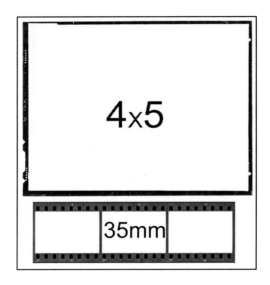

Size comparison between a medium format and 35mm film negative.
Courtesy IStockPhoto.com/ Winston Davidian

Medium format press camera, circa 1940.

Courtesy IStockPhoto / Natalia Bratslavsk

The Digital Revolution

Just as advances in science allowed for the progression from the camera obscura to the view cameras of the 1800s, to the 35mm, medium, and large formats of the twentieth century, further advances in science have taken photography to a frontier that would seem unheard of just a quarter century ago – a purely digital environment.

With the advent of digital photography for the consumer market by the late 1990s, the photographic industry experienced a paradigm shift of monumental proportions.

Initially, technology prevented digital camera equipment from being readily available beyond the journalism industry.

It was not until the dawn of the twenty-first century that technology on both sides of the film versus digital divide within the photographic industry equalized, meaning that digital camera equipment was finally able to match the resolution capabilities of their traditional film counterparts.[10]

With the most current data, photographers have argued that once digital cameras reached a resolution of approximately six megapixels (approximately six million pixels), the human eye was unable to discern a print created digitally from a print created using traditional developing techniques.[11] As a point of reference, major camera manufacturers first shipped consumer-grade digital cameras capable of this resolution to the North American continent in 2003.

Photographically speaking, it is important to note that there is a threshold beyond which the differences between analog (film) and digital photography are so slight as to be imperceptible to the human eye.

This becomes important because, physically speaking, photographic equipment has changed very little from the camera obscura detailed by Leonardo da Vinci in the fifteenth century and a digital camera in use in America today.[12]

Nikon D70 digital camera
Courtesy Marie-Lan Nguyen

(Endnotes)

[1] http://en.wikipedia.org/wiki/Leonardo_Da_Vinci

[2] http://en.wikipedia.or a g/wiki/Leonardo_Da_Vinci_-scientist_and_ inventor#Leonardo.s_journals

[3] (Capra, 2007)

[4] (Capra, 2007)

[5] http://en.wikipedia.org/wiki/Codex_Atlanticus

[6] http://en.wikipedia.org/wiki/Ibn_al-Haytham

[7] http://en.wikipedia.org/wiki/Photography

[8] http://www.rleggat.com/photohistory/history/daguerr.htm

[9] http://www.rleggat.com/photohistory/history/niepce.htm

[10] http://en.wikipedia.org/wiki/Digital_camera#History

[11] http://www.cambridgeincolour.com/tutorials/digital-camera-pixel.htm

[12] (Roesler, 2006)

2. Photography as a Tool in Unexplained Phenomena Research

orensic photography techniques similar to those used in the law enforcement field are also employed in unexplained phenomena research when an investigator (or more commonly, an investigative team) does an on-site field investigation and personal interview with a witness or experiencer.

Forensic photography in this type of setting is fairly commonplace, regardless of whether or not the investigator in question has previous law enforcement training. A photograph taken during an initial investigation can be considered of forensic value simply by including the object in question and enough detail of the surrounding area to be used in later evaluation by experts.

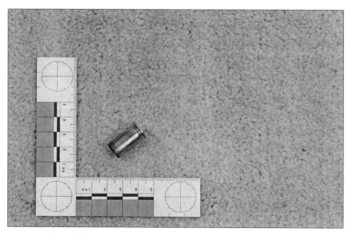

A square, such as this one, is often used in forensic photography to show scale.
Courtesy IStockPhoto / James Ferrie

Forensic or crime scene investigation standards, or, at the very least, reasonably similar standards, can easily be applied to investigation of unexplained phenomena. There is, fundamentally speaking, no difference between photography done in a forensic setting, for example at a crime scene for later use in a court proceeding, and photography that is done during a UFO field investigation. In both cases, the results are images that are taken in a systematic manner, conditions documented, and later scrutinized under a prescribed set of scientific criteria before finally being submitted for expert or peer review.

The tools needed to conduct a forensic quality UFO field investigation are, often times, a combination of the simple and the high-tech. For example, the tools one might carry on a "typical" investigation might include notebooks, legal pads, or pre-printed forms for taking witness statements, making personal notes on the investigation, or noting questions to be asked in a potential follow-up interview, a disposable 35mm or handheld digital camera to document the scene, as well as a digital recorder or small cassette recorder to record witness statements or the investigation itself, or quite simply, the investigator's own personal observations for review at a later time, when an official report is written.

It is for this reason that photography can prove such a valuable tool in unexplained phenomena research, such as the research conducted into the UFO topic. The use of photography in this field can be two-fold; the case investigation itself may center on an anomalous object that was captured in a photograph by a citizen witness, or the investigator (investigative team) may take a series of photographs to augment their initial field investigation, such as has become common practice for many investigators in the field of ufology.

From an investigative standpoint, the main objective is to document the scene, observing everything while disturbing, under ideal circumstances, nothing.

This is of the utmost importance when you are dealing with private property, such as an individual's home. As members of all-volunteer organizations, researchers into paranormal phenomena are operating as private citizens, and, as such, are oftentimes open to liability issues when conducting an investigation on property such as a private residence.

Witness-submitted photographs in
UFO sighting reports

With UFO research, like all other endeavors, it seems, becoming more and more dependant on technology, more researchers, myself included, have grown accustomed to receiving case files, either in an official capacity or from colleagues seeking to have their findings vetted by an impartial third party, via email. A majority of these cases involve only a witness narrative or statement involving an incident in question.

In many cases, witnesses are able to provide a detailed statement and, oftentimes, a diagram or sketch of the object that they report having seen. It is a rare occasion; I would estimate a number of about five to ten percent of cases, which have a photograph from a reported UFO witness included in the case file.

In the event that a photograph is present, it is best to gather as many supporting details from the witness narrative as is possible. It is at this point that, hopefully, the witness with have provided the investigative team with enough secondary details to go about formulating a hypothesis as to what possible scenario might fit the facts that are available.

This principle is put into very real-world terms by Edward Sulzbach, [13] former criminal profiler and instructor at the FBI Training Academy in Quantico, Virginia:

"Never look for unicorns until you run out of ponies."

In simple English: do not make suppositions until you exhaust the most prosaic explanations.

Common causes of misidentification

While sightings of unconventional flying objects continue to be reported in all corners of the globe, of all cases reported and investigated, only a small number of those cases can be certified as anomalous after the investigative process is completed. This is not to say that these cases are the result of hoaxing or intentional deception on the part of the reporting witness. To the contrary, the overwhelming majority of individuals who report anomalous experiences genuinely have no alternative explanation or frame of reference by which to explain their experience. Often, UFO case reports that contain accompanying documentary evidence such as photographs, can be explained as being misidentified naturally occurring natural phenomena, such as an errant reflection of light off of a surface somewhere off-camera, a misidentified aerial object, whether that object be manmade, such as a commercial or military aircraft, satellite, or is instead a celestial object, for example, the planet Venus.

Solar effects, such as the one seen here, are responsible for a percentage of UFO reports that stem from misidentification.
Courtesy MUFON IOPedia

Many times, UFO case reports that are the byproduct of simple misidentifications, such as a camera malfunction, light leakage, or a lens flare, are initially apparent, and an experienced investigator can close interest in a case that they believe to be attributed to misidentification by attempting to replicate the questionable or unknown photograph themselves. It is the personal practice of this writer, however, to follow up by consulting another researcher who has photographic analysis experience, such as Dr. Bruce Maccabee, Maryland MUFON State Director, and asking for a professional opinion.

This is roughly the same process that would be adhered to in a criminal proceeding. One investigator's opinion is never enough. It will be called into question. If faced with a questionable item that you, as an investigator, cannot readily identify with certainty, seek out another objective person to either confirm or refute your findings.

Applying the scientific method to UFO cases involving photography

One chief complaint leveled against UFO researchers comes, almost exclusively, from the halls of academia. The claim is made, erroneously, that academics engaged in UFO research are doing science a disservice or are practicing bad science[14] because, it is claimed, UFO investigations cannot hold up to scientific scrutiny. That has shown to be inaccurate, as the "best evidence" UFO cases are the cases that have withstood the test of time and scientific inquiry.[15] [16]

Scientific methodology is at work in UFO investigation on a daily basis, [17] but particularly with regard to UFO cases where photographic images are involved. UFO photographs are not blindly accepted by genuine proof of something extraterrestrial. They are analyzed by experts in optics, with specialties in both still photography and video imagery.[18]

Even then, as with expert forensic testimony in criminal cases, the results of many scientific tests cannot be guaranteed absolutely — only with a high degree of probability.

The use of digital imaging software
in photographic analysis

Advances in photographic technology have also led to advances in other areas of digital imaging. Perhaps the most well-known and widely commercially available piece of digital imaging software is Adobe's *Photoshop* imaging suite. While the professional grade version of this program, common to graphic designers and professional photographers, can carry a retail price tag close to $1,000, more basic versions of the program are available at virtually any office supply store. Some of the features common to *Photoshop*, including the ability to composite multiple images together, remove and replace segments, such as a background, from an image, as well as the ability to alter an image by way of a digital "airbrush" make it a popular source of doctored or otherwise altered photographs.

Given that such sophisticated editing software is so readily available both to amateur and professional photographers, as well as the general public, both the news media and photography trade groups, such as the National Press Photographers' Association, have gone on record to discuss what is and is not to be considered acceptable industry practice, as far as retouching and editing of digital images is concerned.[19] Common, acceptable examples of digital photo editing include adjustments made to the image that do not change that image's news value (i.e.: changing a color image to grayscale, adjusting the base values of an image, such as brightness and contrast), as well as other techniques, such as a technique common in promotional and advertising photography, known as *"spot healing"*, where minor imperfections in an image, such as tattoos, surgical scars, or other visible imperfections may be digitally removed from a promotional image before a piece goes to print.

Due to the fact that photographic images can easily be taken out of context when they are altered in even the minutest way, print and digital media outlets impose stringent controls and codes of ethics with regard to the use of images that have been digitally altered or compromised in any way.

As an example of how altered photographs can enter the public domain, the University of Wisconsin at Madison came under fire in September of 2000, for knowingly distributing an undergraduate admissions application that featured a digitally altered cover photograph. The image was a composite of two separate images of crowds at University of Wisconsin football games, although, it was later revealed that the images used had been taken over a span of two years and composited together.[20]

Such an ethical dilemma is the perfect real-world example of why the news media has lobbied for such a stringent code of professional conduct with regard to the way digital images are used both in print and on the Internet.

The fact that photographs can be so easily manipulated in a digital environment is the primary argument raised by those that would debunk the entire UFO subject as to why UFO reports with a photographic component should be summarily dismissed, as they lack a point of reference to determine factors such as distance and speed[21]. There are others that will dismiss all photographic UFO cases outright, saying that they could be the products of digital forgery.

That argument can only potentially be true in the most modern cases; certainly this cannot be the case where photographs date to the 1960s or earlier.

This being said, the same technology used to create an altered photograph can be used in reverse, to detect inconsistencies in an image, which would suggest that it has been altered in some way, and, as such, is not a genuine photograph.[22]

The prevalence of digital images and UFO-related material available via the Internet are two factors frequently mentioned by those individuals who actively seek to debunk the UFO subject.

While the lack of safeguards on a bulk of material published to the Internet ostensibly puts much of the research material in the public domain and, as such, into the proverbial "court of public opinion", what debunkers of the UFO subject fail to mention is that there have been a large number of UFO sightings, in the past half century alone, that have involved photographic evidence, all of which have withstood scientific scrutiny on multiple occasions.[23]

Photographic evidence as a piece of the overall puzzle

As mentioned previously, photographs are a factor in an exceedingly minute number of cases, compared to the number of cases reported in any given year. It is this writer's opinion that photographic evidence cases represent such a small percentage of cases simply because said cases go underreported, if not unreported entirely.

The lack of reporting what could very well be legitimate anomalous events puts serious, scientific-minded investigators at a distinct disadvantage. They are simply, pardon the pun, not seeing the full picture because they have not been given all the available information.

If the UFO subject were treated in the United States as it is in other countries around the globe, that is to say dealt with openly and as a matter of science, there would be, almost certainly, a marked increase in the number of UFO events reported annually. With an increase in overall reporting, it would stand to reason that the number of anomalous photographic cases reported would see a dramatic increase as well.

(Endnotes)

[13] (Cornwell, 2002)

[14] (Roesler, The Invisible College: UFO Field Research in Academia, 2007)

[15] (Roesler, The Invisible College: UFO Field Research in Academia, 2007)

[16] (Swiatek, 2005)

[17] (Jacobs, 2002)

[18] (Sainio, 2002)

[19] (Roesler, Digital Wizardry in the UFO Age, 2006)

[20] http://www.cs.dartmouth.edu/farid/research/digitaltampering/

[21] (Shostak, 2005)

[22] (Roesler, "We Are Not Alone" [radio program]. Dearborn, MI: WHFR-FM, 2007)

[23] (Maccabee, 2005)

3. The Forensic Value of UFO Photographs

P hotographs taken during suspected UFO sightings can quite easily be sources of forensic data. While it might seem implausible to those not familiar with the subject, forensic data can reasonably be found in photographs taken during the sighting of an unconventional flying object.

While the object in question may not be initially identifiable, other factors present in an image may be able to be scientifically measured, and using mathematics, a degree of scientific information may be obtained.

Consider the following example:

An investigator receives an original photograph from a witness during the course of their investigation. This photograph is taken in the early evening hours, at the approximate time of sunset. The witness is able to photograph an anomalous object in the sky across the street from their home. This object appears, in the witness's estimation, to be some five to ten feet above the streetlight across from their home.

Commonly, an investigator might ask a witness what they perceived the approximate size of the object to be, in reference to an object, such as a penny, held at arm's length. This type of questioning can provide a reasonable reference point, with respect to scale. This data can be incredibly important, as it is used to determine the size of the object in question.

For example, there have been cases where unidentified objects or aircraft have been reported by witnesses and, during later investigation, deemed to be larger than both commercial and military aircraft.

It is for this reason that it is advantageous to, if at all possible, include landmarks or other large stationary structures, such as street lamps, buildings, or the roofs of houses in photographs. These types of structures can later be used by investigative personnel to determine distance, altitude, and other factors that might aid in identification.

Commonalities between UFO photographs and crime scene photographs

There are commonalities, while not necessarily immediately apparent, between UFO photographs and crime scene photographs. While crime scene photographs represent the aftermath of an incident, photographs which accompany UFO investigations are taken in real time.

Commonalities exist between both sets of photos and their evidentiary value. Both sets of photographs tell two separate stories. They tell their stories in images, as well as telling a much more detailed, if abstract, story hidden within the details that make up the image.

The saying goes that *"a picture is worth a thousand words"*, and this very concept has been shown to be true time and again, when technical analysis reveals details not apparent upon initial inspection.

For example, a photograph of an automobile accident might reveal details as to time of day, weather conditions, and temperature, based on subtle, if not otherwise imperceptible details such as gradation of tint in window glass, angle of reflection, and lighting.

Photographs as a source
of secondary case information

The information contained in any photograph can extend far beyond the visible spectrum. A trained eye and a reasonable amount of insight can tell the investigator a great deal about that single moment in time when any particular photograph was taken.
The most minute details present in any given photograph can give any researcher insights into details that the witness in a case either would not have thought of, was not consciously paying attention to, or had no way of knowing.

Such is the case with the 1965 Santa Ana, California photographs – commonly known in the field of ufology as The Heflin Photos. Analysis done on the photographs in the mid-1990s confirmed Heflin's approximations as to altitude and distance, while revealing details as to the speed and velocity of the object, which had previously gone undetected in earlier investigation.

It is these types of details that commonly go unnoticed or undetected altogether during an initial investigation, but might later become apparent through either the advancement of technology or the advent of new technologies altogether.

Types of information that can be obtained
through photographic evidence

As mentioned previously, photographs can serve as a source of secondary information in any given case. In a manner of speaking, what a photograph does not show is almost as telling as what is visible. Based on angles, lighting effects, and atmospheric conditions, among other examples, it is possible to discern additional information about the circumstances under which a photograph was taken.

For example, based on the position of the suspected UFO (unconventional flying object) in a photograph in relation to other objects in the frame, details such as altitude, distance from camera, and even the approximate size of the object can be determined within a reasonable degree of accuracy.

Further information can be gathered by taking into account the physical characteristics of the camera equipment used to capture the images in question. By studying the technical specifications of any given camera, plus the capabilities of the lens, reasonable estimations can be made, again, as to altitude, speed, direction, and the physical dimensions of the questioned object.

The January 2, 1975 photograph captured by Michael Lindstrom in Kauai, Hawaii, bears out just such a methodology, as documented by a veteran UFO researcher Rob Swiatek in *UFOs Exposed: The Classic Photographs*[24]. The photographer reported to viewing the object for approximately two minutes, snapping three color slides with a Pentax camera with a 135 mm lens.

Later, calculations made by Dr. Bruce Maccabee, an optical physicist, reveal that if the witness's initial estimate as to five thousand feet altitude is correct, the object would have measured some thirty-three feet in width, and the maximum slant range from photographer to object would have been approximately 18,000 feet.

Chain of custody as it relates to photographic evidence in UFO cases

As in any investigation, chain of custody is a vital part of the evidence collection process. Without an established chain of custody, a case based largely on circumstantial evidence would begin to unravel.

The term chain of custody refers to documenting each step in the evidentiary process, from the moment a piece of evidence is collected, in this case a photograph, until the conclusion of an investigation.

This is simply one point of congruence among many between law enforcement investigation and the investigation of unconventional flying objects or UFOs. As with any investigation, documentation of each and every step of the investigative process, no matter how small, is absolutely essential. A saying common to law enforcement personnel with regard to this practice is *"If it isn't written down, it didn't happen"*. As the quote suggests, if you do not have a written record of your actions with regard to a particular investigation, you have no way to document that you in fact conducted yourself according to policies and procedures. This is particularly important when, during the course of an investigation, a witness turns over to an investigator any item of personal property, such as a questioned photograph. By documenting chain of custody, it is possible to relieve yourself of any unnecessary level of liability, due to any unforeseen circumstances.

Chain of custody documentation allows anyone reading a case file to know, at a glance, exactly how many individuals have handled or seen any individual piece of evidence throughout the course of an investigation.

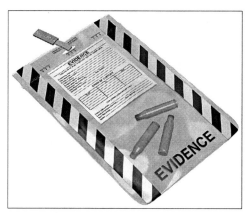

An evidence bag, such as this one, helps to document chain of custody in criminal investigations.

Courtesy IStockPhoto / Stephen Sweet

One method of documenting chain of custody is with a standardized form meant to be used specifically by the individual researcher and not necessarily for public disclosure. While these forms may become public record due to their inclusion in a case file, this is a reasonable way of clearly and concisely keeping all pertinent case material within reach for future reference. This can be especially useful during an investigation which involves either multiple witnesses or multiple members of an investigative team.

Specifically with respect to photographic evidence, chain of custody, especially where secondary analysis may be needed, can be established quite simply. In many cases, this process is as simple as keeping thoroughly detailed notes and copies of all incoming and outgoing e-mail messages with respect to the case being investigated. If, for example, the original case investigator has reason or need to turn over a piece of evidence for additional study or analysis, this should be carefully noted including the name of the person to whom the item was given, contact information, as well as the date and time of the transfer.

It is important to note that, if at all possible, when dealing with photographs, it is advisable to send copies of photographs rather than original images, due to the possibility of loss or damage. The notable exceptions to this rule are cases which involve digital images. Digital images, while susceptible to editing and digital manipulation, can simply be forwarded as e-mail attachments with minimal risk of damage or loss.

Outside authentication of anomalous photographic images

During the course of an extended UFO case investigation, where photographic images of questionable, suspect, or otherwise indeterminate origin exist, the initial investigator must, invariably, determine to the best of their ability, the cause of the anomalies captured in the photograph.

Often, an investigator's first stop might be contacting the lab that processed the roll of film from which the questioned photographs originated. Contacting the photo lab might be all that is necessary to resolve the case. If a lab technician notices similar anomalies to those shown in the questioned photographs on prints belonging

to other customers, the cause of the anomalous photograph could be ruled as a processing error due to light leakage during print development.

If, in contacting the photo lab, the investigator finds no discernible evidence to suggest an error in processing, yet wishes and expert opinion, many commercial film manufacturers, such as the Eastman Kodak Company in Rochester, New York, will ask to examine the questioned print, often at no charge to the consumer.

This process frequently examines prints, negatives, and in some cases, even entire commercial labs.

During a manufacturer-initiated check, photographs and negatives are routinely checked to ensure that the questioned image did not result from a manufacturer or processing error, such as faulty cutting of negatives, printing from the wrong side of a negative, resulting in an inverse image, or misprinting of the image in its entirety, commonly as a result of printing equipment and being improperly set.

Regardless of the findings of lab analysis of a photograph, it is standard practice for said lab to issue the customer a detailed report, delineating exactly what tests were performed and the results of each.

If a photograph containing an anomalous aerial object is submitted to this type of testing and passes a lab inspection with no identifiable cause, the lab report immediately becomes a very valuable piece of secondary evidence. In the event that a UFO investigation ultimately lead to a courtroom setting, this type of scientific analysis becomes invaluable.

Most frequently, however, UFO investigations that involve photographic images make their way from the initial investigator to an independent specialist, such as Dr. Bruce Maccabee or Jeffrey Sainio, both of the Mutual UFO Network.

It is this third-party review process that, often times, either definitively identifies the questioned object in a suspect photograph or, in rare cases, classifies the object is unidentified.

Although, to many, photographs of an unidentified or as yet unidentifiable object in the sky are the things of science fiction, it is the work of experts in the fields of optics and physics, among others, that provide scientific credibility to seemingly incredible events.

(Endnotes)

[24](Swiatek, 2004)

4. The Battle of Los Angeles

The story of the *'Battle of Los Angeles'* UFO case is truly the story of a world at war. I say this due to the fact that this case, perhaps one of the most incredible and least well-known UFO sightings of the modern era, occurred some sixty days following the attack on Pearl Harbor, Hawaii and the entrance of the United States into the Second World War.

To put this incident into the proper prospective, it is important to note the date of the event – Wednesday, February 25, 1942 – a full five years before the world would hear of pilot Kenneth Arnold, 'flying saucers', or Roswell, New Mexico.

Kenneth Arnold, the pilot who coined the term 'flying saucer', circa 1947.
Courtesy Black Vault Encyclopedia Project

Under normal circumstances, the sighting of a strange aircraft over one of America's largest cities in view of multiple witnesses, especially in the dead of night, would be enough to cause alarm. In this case, however, that was only the beginning.

The *Battle of Los Angeles* has its roots in an earlier and otherwise unconnected event. On the preceding Monday, February 23, a Japanese submarine had surfaced and fired upon the Ellwood oil field, located some twelve miles northwest of Santa Barbara, California, causing a surprisingly small amount of damage, estimated in news reports at five hundred dollars. This is, in itself, remarkable, considering that over 200 pounds of shell fragments were recovered from the Santa Barbara attack.

Given the Japanese air attack on the US Naval installation at Pearl Harbor, Hawaii that took place in December of 1941, it is easy to view the events of that winter, from Pearl Harbor onward, as one series of events, as the residents of Los Angeles must have undoubtedly done on the night of February 25.

This being said, there is no direct evidence at present to link the Japanese submarine incident at Santa Barbara on February 23 with the aerial encounter over Los Angeles forty-eight hours later.

Unidentified craft sighted over Culver City, California

The event which has come to be known as *The Battle of Los Angeles* began as an unknown aircraft, thought to be an invading air force from Japan, entered the airspace over Culver City and Santa Monica, California, just outside of the City of Los Angeles proper.

One of the elements of this case, which made it all the more sensational, was the fact, as stated previously, that these events occurred in barely sixty days after the bombing of Pearl Harbor, Hawaii by the Japanese and, subsequently, the entry of the United States into the Second World War. As a result of the events at Pearl Harbor, there was obviously a heightened sense of anxiety by the residents of Los Angeles, which likely led to the links made, whether legitimate or imagined, between the surfacing of a Japanese submarine off Santa Barbara on February 23 and the sighting of this unidentified craft in Los Angeles on February 25.

The incident was additionally dramatic due to the fact that it occurred in the very early morning hours of February 25. The City of Los Angeles was, at the time, dealing with a blackout which stretched from the city itself, including several landmarks on the west side, including MGM Studios in Culver City, to the Mexican border in the south to the San Joaquin Valley.[25]

The forty minute air war

Perhaps the most notable aspect of the Los Angeles case is the fact that the object – whatever it may have been – was not only repeatedly photographed, but was brazenly fired upon by dozens of the U.S. Army anti-aircraft batteries in full view of Los Angelinos for nearly forty minutes.

There are many discrepancies in various accounts of the incident, including the time and duration of the incident itself.

One report notes a duration of thirty-eight minutes, from 3:36 AM to 4:14 AM, while another account of the incident cites a completely different time span, from 2:21 AM to 3:06 AM, a span of forty-five minutes.

While only one account can be correct, as far as time, the exact time of the event may be immaterial in comparison to the events reported.

Even taking discrepancies of time into account, the witness accounts and news reports all report the same basic facts. The mysterious craft hovering over Los Angeles was fired upon by the U.S. military in a one-sided confrontation that saw the military expend some 1,440 rounds of anti-aircraft munitions against an unknown object whose speed varied from a near-crawl to an estimated two hundred miles an hour.

Additional reports of the event report there being dozens of smaller "planes" or smaller craft seen during the course of the event, as well as the presumably larger craft which seems to be the primary focus of the event. [26]

Los Angeles Times, February 26, 1942
Courtesy Black Vault Encyclopedia Project

The Battle of Los Angeles as a precursor to modern UFO events

The 'Battle of Los Angeles' UFO incident, without question, served as a precursor to modern UFO cases. This can be said due to the simple fact that it predates the Kenneth Arnold sighting and Roswell Incident by some five years. While this is true, the Battle of Los Angeles incident can certainly be said to foreshadow a number of UFO events in the modern era.

Particularly curious about the Los Angeles event is the memo drafted by General George Marshall to President Franklin Delano Roosevelt on this matter, which remained classified in the interests of national security until 1974.

As author and researcher Richard Dolan notes in his mammoth *UFOs and the National Security State*[27], the military explanations were inaccurate at best, bald-faced lies at the worst. Dolan specifically notes the fact that the Secretary of the Navy, Frank Knox, went so far as to chalk up the entire Los Angeles incident to "war nerves", claiming that there had never been any aircraft other than those belonging to the US military in the airspace over the City of Los Angeles.

The local press took particular offense to the Navy's suggestion, with one local newspaper, the *Long Beach Independent*, going on record saying *"There is a mysterious reticence about the whole affair and it appears some form of censorship is trying to halt discussion of the matter."*

As researchers such as Dolan and John Greenewald can attest, it is telling, though not particularly surprising, that in the thirty-two years between the event and the release of the Marshall memo, the Department of Defense claimed to have no records pertaining to the Los Angeles event.

Dolan puts a very fine point on the whole affair, noting, in *UFOs and the National Security State*, that *"five years before Roswell, the military was already learning to clamp down on UFOs."*[28]

The Los Angeles sighting can be viewed as a precursor, in some respects, to, specifically, the Phoenix Lights case of 1997 and the many cases throughout the years where there was military interaction with an unknown craft, both by the United States military or those of foreign governments.

While the Los Angeles event foreshadows the 1997 Phoenix case primarily due to the fact that both occurred at night over major population centers, there are many modern cases that fit these parameters, just as there are multiple cases where military interaction has been a component, as mentioned, both on foreign soil and within the borders of the United States, though not all of these cases necessarily are supported by documentary photographic or video evidence.

Analysis of the 'Battle of Los Angeles' photographs

The primary, and, to my knowledge, only photographic analysis performed on the photograph shot during the Battle of Los Angeles was performed by Dr. Bruce S. Maccabee, United States Navy optical physicist and the Mutual UFO Network's State Director for Maryland.

To better understand the core truth, if you will, of this incident, it is important to look at all of the surrounding information, much the same way that all areas of an image are taken into account during photo analysis.

As a photograph is a single moment frozen in time, I believe it is important to include the subsequent news coverage of the event, both by the *Los Angeles Times* and other publications of the day, as part of the investigative record. This is as close to eyewitness testimony as we will find in this particular case

The caption which appeared beneath the photograph in the February 26, 1942 *Los Angeles Times*, alongside the report of the Los Angeles blackout, anti-aircraft shelling, and aftermath reads:

"SEEKING OUT OBJECT – Scores of searchlights built a wigwam of light beams over Los Angeles early yesterday morning during the alarm. This picture was taken during blackout; shows nine beams converging on an object in sky in Culver City area. The blobs of light which show at apex of beam angles were made by anti-aircraft shells."

This caption, upon a more critical inspection, gives away some more telling photographic clues than are initially apparent.

The fact that nine separate beams of light are visible in a single frame of film, in conjunction with that fact that multiple explosions are also present, offers some information as to the likely settings of the camera when this photo was shot.

As Maccabee posits in his analysis:

> *I don't know the film speed or the f stop of the camera. However, I would guess that the f-stop was low (lens "wide open"; f/2 or 3?) and that this is a time exposure because (a) the light beams show up and (b) there are quite a few "explosions" (I presume)which probably did not happen all at once. The exposure could have been several seconds.*
>
> *The fact that the beams basically do not get past the "object" (there is some faint evidence of beams above the*

object), whatever was at the beam convergence must have been optically quite dense. If there was a lot of smoke swirling around the volume of air illuminated by the beams, I would expect to see variations in beam brightness (brighter where there was smoke). There are variations, but they are uniform and agree with the distance (from the searchlight) and width of the beam. That is, the variations are consistent with each beam getting dimmer as it travels away from the searchlight.[29]

I will deal with Dr. Maccabee's conclusions, which I find to be quite astute, on a point-by-point basis. In my estimation, Dr. Maccabee's observations have a high likelihood of being correct.

As we do not know the film speed or the f-stop settings of the camera which shot this photograph, we can only theorize. This being said, the negative image below, and the corresponding positive print both suggest that the scenario outlined by Dr. Maccabee, with a very low f-stop, such as 2 or 3, is highly likely. A low f-stop would allow the camera aperture to remain open for an extended period of time, most likely several seconds, although any statement as to exactly how long would be pure supposition on the part of this writer.

The multiple beams from the searchlights and multiple shell explosions present in the frame also go a long way toward suggesting that this is a timed exposure, as searchlights sweep at a fixed rate across the sky. The relative brightness of the original image in itself suggests a time-lapse exposure. This event occurred well past midnight, and the photograph which ran in the *Los Angeles Times* was illuminated in large part, if not exclusively, by the light given off from the searchlights present at the scene, as well as the explosions occurring overhead.

The only conclusion that, to this writer, seems logical is to concur with Dr. Maccabee's independent analysis. The photograph in question could not have been doctored by any conventional means of the day, and more importantly, what the photo shows is consistent with the reports of the event. It is curious that the unidentified object appears to be so dense that its shape can't be determined, even with the searchlights and anti-aircraft batteries in the vicinity.

One could argue, however, that any depiction of the object could be inaccurate due to its movement over the duration of the time-lapse photo.

Negative Image of the Los Angeles Times cover photo.
Courtesy Black Vault Encyclopedia Project

What we learn from the Battle of Los Angeles

The Battle of Los Angeles, like all unresolved UFO case reports, invariably raises more new questions than it answers. Most certainly, we will never know for sure what was seen over Los Angeles on the morning of February 25, 1942.

Judging from witness reports and additional documentary data, including the Marshall memo and Los Angeles Times photographs, one can reasonably assume that what was reported was not a conventional aircraft.

The evidence surrounding the military shelling that took place in conjunction with the sighting seems to confirm this hypothesis.

There is, at present, no prosaic explanation that would fit the fact that any aircraft was bombarded by approximately two thousand rounds of twelve-pound high explosive shells for nearly forty minutes and was not shot down.

A very persuasive argument for the Los Angeles craft being some form of advanced technology can be made simply by noting the collateral damage suffered during the military's shelling. The drama played out over Los Angeles would result in the deaths of six people as well as property damage to city streets, public buildings, and private homes for miles around the sighting location.

While the whole truth surrounding the Los Angeles event will likely never be known, what information is available does leave unanswered questions, but does leave interested parties with a frame of reference from which to draw educated conclusions.

The Battle of Los Angeles is arguably one of the most incredible mass witness cases in modern UFO history. Like the 1952 Washington, DC sightings, the Los Angeles case is another case which occurred in a major metropolitan area for an extended period of time, which included a fairly large number of eyewitnesses.

While these factors don't definitively prove the hypothesis that the craft fired upon in Los Angeles was in fact extraterrestrial in origin, this case does alter the generally accepted timeline of modern ufology, as it occurred some five years prior to the Roswell, New Mexico event.

This case also exhibited characteristics later reported time and again within the field of ufology, particularly the fact that military weapons could not shoot down this craft. There are multiple reported sighting cases on record involving military and commercial aircraft and personnel. As the United States military's interaction in this particular case is central to the narrative, the Battle of Los Angeles case becomes particularly valuable when placed in historical context.

As stated previously, unresolved UFO cases, by their very nature, present unique unanswered questions. Perhaps early cases in ufology, such as the Battle of Los Angeles, will provide telling details only when viewed in concert with much later events.

(Endnotes)

[25] (*Los Angeles Times*, 2/26/1942)

[26] www.sfmuseum.org/hist9/aaf2.html

[27] (Dolan, 2002)

[28] (Dolan, 2002)

[29] http://brumac.8k.com/BATTLEOFLA/BOLA1.html

5. The Roswell Incident

Roswell, New Mexico. Just saying the name conjures up images of wide expanses of desert, a crash site, alien beings, the U.S. military, and a cover-up in the minds of most Americans, whether they are students of ufology or not. Some of the publicity is unfair and sensationalized, thanks in large part to countless television documentaries on the case, many of them touching only briefly on the facts of the case, focusing attention instead on drawing ratings, rather than providing much of educational value.

The slippery slope, it seems, with respect to the Roswell Incident, as with other cases of an earlier vintage, is that many firsthand witnesses have since died, information has been lost or long forgotten, leaving secondhand information, anecdotal data, conjecture, supposition, and educated inference as the "best evidence".

This is not to say that there has not been serious, substantive research done into the Roswell Incident that has advanced the debate significantly. There has. Retired nuclear physicist Stanton T. Friedman, considered by many in the UFO research field to be the preeminent authority on the Roswell Incident, has studied this case almost exclusively for nearly forty years, lecturing on the Roswell Incident and the supporting military evidence the world over.

There have been entire books written concerning the events at Roswell and Corona, New Mexico. Of this rather robust catalog, the work generally considered to be the definitive account of the

Roswell matter is *Crash at Corona* by Stanton Friedman and Don Berliner.

In the interest of brevity, I will offer an overview of the Roswell Incident, provide what is generally accepted as the chronology of the Incident and its aftermath, and focus specifically on what I consider to be the forensic evidence of the Roswell Incident and subsequent U.S. military investigation.

There are those who will debate aggressively whether or not a crash actually occurred at Roswell, New Mexico. There are reports of a similar event occurring at Corona, New Mexico, with Roswell being the nearest population center and the base of operations for the US Army Air Force's 509th Bomb Group.   It is likely that Roswell was the center of the investigation due to the military's base of operations, regardless of whether the crash which the military was charged with investigating occurred in Roswell proper or in the surrounding area. In keeping with the commonly accepted history of the event, when referring to the events of July, 1947, I will refer to them as "The Roswell Incident".

When the Roswell narrative is retold in the mainstream media, it seems as though the narrative begins with the entry of military personnel onto the scene. It is important, as with any narrative, to get a full and complete picture of the events in question. It is the personal opinion of this writer that the Roswell Incident suffers a great disservice at the hands of the general public when the complete sequence of events is not explained in detail.   While this account is by no means meant to be an exhaustive retelling of this seminal case in the history of ufology, I trust that it serves both as a reasonable primer on this case, as well as a fair and balanced argument for the inclusion of the Roswell Incident alongside cases where photographic evidence is a chief component of the case.

The Army Air Force mounts a recovery operation

When strange pieces of metallic debris appeared on the Foster ranch, rancher W.W. "Mac" Brazel collected a few pieces and took the material to the most logical person, in Brazel's mind, in this small community – the local sheriff, George Wilcox.

It was at the suggestion of Sheriff Wilcox that Brazel brought this material to the attention of military personnel based at Roswell Army Air Field, home of the 509th Bomb Group. As the 509th Bomb Group was the home of the *Enola Gay*, the aircraft responsible for dropping the nuclear payload on Hiroshima and Nagasaki during the Second World War, both the sheriff and Brazel reasoned that the unknown debris would be easily identifiable to the personnel of the 509th.

When discussing the Roswell Incident, it is important, if not vital, to also discuss the man who has become synonymous with the case – U.S. Army intelligence officer Maj. Jesse Marcel.

Marcel, a decorated military officer of impeccable character, headed the recovery operation. It is important to note Marcel's character when discussing the events surrounding the Roswell Incident because, for one reason or another, Marcel, as head of the on-site investigation and recovery operation in New Mexico, has become the target, even posthumously, of personal attacks by those attempting the disprove, discount, or discredit the Roswell events.

Perhaps no one describes this better than Marcel's own son, Jesse Jr., in his 2008 book, *The Roswell Legacy*.

Maj. Jesse Marcel, circa 1947
(Courtesy Black Vault Encyclopedia Project)

Dr. Marcel, himself a decorated military veteran describes in great detail, the career, both pre-and post-Roswell, of his father. Dr. Jesse Marcel, Jr. extols his father not simply as a proud son, but to underscore the point that his father was not the type of man or soldier prone to flights of fancy. Dr. Marcel notes, in *The Roswell Legacy* that

> *"He was thought of very highly by his superior officers, both before and after the Roswell event. His marks are generally excellent with an overall rating of high excellence. One report from one individual had him as unimaginative, but I would think that would give him more credibility in describing the debris: If he was not imaginative, how could he have imagined debris from a weather balloon as having come from a flying saucer?"*

According to the commonly accepted chronology of events, as well as statements on record in various books written on the case, such as *The Roswell Incident,* published in 1980, as well as *UFO Crash at Roswell (1991),* Major Marcel's on site investigation occurred on Sunday, July 6, 1947, with Marcel and other officers returning to Roswell the following day.

As the entire sequence of events transpired in what seems to be quite a rush, this fact alone would seem to suggest that the object which crashed and was recovered in the desert outside Roswell was anything but conventional. Maj. Marcel returned to Roswell on the evening of Monday, July 7. He would find himself on a plane bound for Fort Worth, Texas, summoned to the office of Gen. Roger Ramey, the following day.

The Roswell Incident goes national

"Headline Edition, July 8, 1947. The Army Air Forces has announced that a flying disk has been found and is now in possession of the Army. Army officers say the missile, found sometime last week, has been inspected at Roswell, New Mexico, and sent to Wright Field, Ohio for further inspection."

This was the report of the Roswell Incident, as it originally went out over the news wire in July of 1947. The exact date of the Roswell crash remains a matter of debate, with the commonly accepted timeline placing the date of the actual crash on or around July 1 or July 2, 1947.

According to a search of *TimeAndDate.com,* July 8, 1947 was a Tuesday, which would put the recovery of a crashed vehicle from the Roswell site *"sometime last week"* between Monday, June 30 and Friday, July 4, 1947, well within the commonly accepted chronology of events.

Roswell (NM) Daily Record, July 8, 1947.
(Courtesy Black Vault Encyclopedia Project)

Calendar for July 1947

Sun	Mon	Tue	Wed	Thu	Fri	Sat
		1	2	3	4	5
6	7	8	9	10	11	12
13	14	15	16	17	18	19
20	21	22	23	24	25	26
27	28	29	30	31		

Calendar generated on www.timeanddate.com/calendar

The Roswell Incident as a photo evidence case

The Roswell Incident can be, and I believe should be, considered a forensic evidence case in the most modern sense, based upon photographs shot by J. Bond Johnson of the *Fort Worth Star-Telegram,* of materials photographed in the Fort Worth office of Brigadier General Roger Ramey, then the commanding officer of the Eighth Air Force, allegedly recovered from the Roswell site, before reaching the final destination of Wright Field, Ohio. (Wright Field is now Wright-Patterson AFB.) The single most famous photograph taken by Johnson is the following, and the reader should pay particular attention to the paper in General Ramey's hand.

This photograph of Ramey and Dubose is perhaps the single most solid piece of forensic evidence in the history of ufology, despite the fact that the materials photographed in the General's office were not the same materials described by Marcel in his initial on-site investigation in New Mexico.

As mentioned previously, it is important to pay attention to the document in General Roger Ramey's hand. In the first photograph taken by Johnson, text is clearly visible. In subsequent photos, the page in Ramey's hand appears blank.

This inconsistency could mean absolutely nothing or it could be the crux of a case for cover-up and misinformation, depending on one's point of view. It is the opinion of this writer that the discrepancy between photos in the Bond Johnson series was the result of, for want of a better term, "over-preparation" on Ramey's part. This is the same reason, in principle, that many career criminals are arrested for the crimes that ultimately land them in prison – no matter how intricate and well organized their methods, there is always one minute detail that is overlooked, and it is this minute detail that proves to be their undoing.

In the matter of the Roswell Incident, I believe this to certainly be true of Roger Ramey.

While General Ramey took great care to keep the contents of the memorandum in his hand out of the direct line-of-sight of the camera and while he certainly could not have foreseen the advances in the fields of optics, photography, and computer science that would take a fairly standard press photo op and turn it instead into the focal point of an investigation that would span in excess of six decades, as of this writing.

The photographs taken in General Ramey's office represented, at the time, the end of the Roswell story.

Presuming that the initial press reports out of Roswell were true, the photos of Ramey, Dubose, Marcel, and the reflective weather balloon material represent political damage control of the highest order. It would be decades before technology would advance to the point where it would allow interested investigators to take a fresh look at this cold case.

Maj. Jesse Marcel, with weather balloon material

at Fort Worth, Texas, July 8, 1947.

(Courtesy Fort Worth Star-Telegram Collection, Special Collections, University of Texas at Arlington Library, Arlington, Texas)

General Roger Ramey (kneeling) and then-Colonel Thomas
Jefferson Dubose (seated) at Fort Worth, Texas, July 8, 1947.
*(Courtesy Fort Worth Star-Telegram Collection, Special Collections,
University of Texas at Arlington Library, Arlington, Texas)*

Technical analysis of the Ramey Memo/J. Bond Johnson Photographs

The Roswell Incident might well have been lost to the pages of history as inconsistency after inconsistency was explained away by the US Army and, later, the US Air Force, were it not for two small pieces of physical evidence – the telegram General Roger Ramey received while in Fort Worth, Texas and a photograph clearly placing it in his hand on the afternoon of Tuesday, July 8, 1947.

Alone, these facts mean nothing – except for the fact that technology has advanced to the point where the photographs taken that July day are archived for posterity – in a digital format – at the University of Texas at Arlington, as well as computer technology which allowed interested parties, specifically researchers from the Mutual UFO Network, to digitally examine (but not alter) the image, so that the text of the memo, some sixty years old, is now readable. While it might be seen as rare in the field of ufology, the digital analysis conducted on the Ramey photograph does conform to the scientific protocols employed in other forensic disciplines, and therefore, would meet the standards generally accepted in the law enforcement and criminal justice fields.

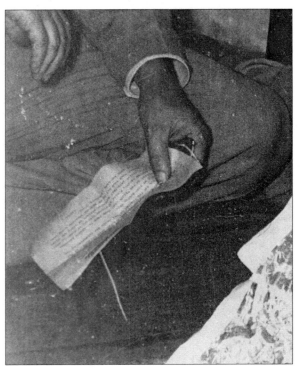

Enlargement of the Ramey-DuBose photograph, showing the "Ramey Memo" in detail.
(Courtesy Fort Worth Star-Telegram Collection, Special Collections, University of Texas at Arlington Library, Arlington, Texas)

It is important to note that while the image may have been manipulated (as far as brightness, contrast, etc.), it was not altered, thereby preserving the evidentiary value of the original photograph. The initial image values have been enhanced, not altered in any way.

The following text, deciphered by researchers using a high-resolution enlargement of the Ramey photograph, was taken verbatim from the November 1998 issue of the *MUFON UFO Journal*, and, despite the passage of time, is still, to the knowledge of this writer, the most true and correct interpretation available to date:

Cradle Telephone or Liberty Bell Symbol >> [Large Underlined Header]
[Top Left] > [Official Crest] handwritten >numerals 15 > 33 > time of receipt?

1) AS THE >??
2) 4 HRS THE VICTIMS OF.THE.YOU FORWARDED TO THE > >
3) AT FORT WORTH, TEX. > >
4) THE "DISK" ?> L AT 0984 ACKNOWLEDGES > > . >
5) EMERGENCY POWERS ARE NEEDED SITE TWO S.W. MAGDALENA, > N.MEX
6) **D** SAFE TALK FOR MEANING OF STORY AND > MISSION.
7) BALLOON STORY. SHOW "STUFF" OF WEATHER BALLOONS SENT ON THE >
8) **** AND LAND L****VER CREWS. > >
9) [blank] > >
10) TEMPLE> >

What the Ramey Memo/J. Bond Johnson photos teach us

Thanks to diligent work by a number of researchers, specifically Dr. Donald Burleson, MUFON State Director for New Mexico, the Ramey memo has become a lasting and viable piece of forensic evidence with respect to the Roswell Incident.

Dr. Burleson, who holds Master's degrees in both Mathematics and English, was the primary researcher involved in deciphering the sixty year old memo from a photographic reprint obtained from the archives of the University of Texas at Arlington.

While the techniques used by Dr. Burleson in the forensic reconstruction of the Ramey memo are fairly technical, they are, in essence, the digital equivalent of colored filters used on a photographic enlarger to enhance a negative image during the printing process.

For example, some common uses of photographic filters are to cut ultraviolet light, correct color balance, or enhance contrast.

62mm ultraviolet, polarizing, and florescent lens filters
(Courtesy Rich Niewiroski Jr.)

In the opinion of this writer, it is particularly interesting to view Burleson's renderings of the Ramey memo, originally rendered using individual colors (magenta, cyan, yellow, etc.), to achieve digitally the same effect that could be produced in a photographic darkroom.

As the color-rendered digital images closely approximate photographic filters, it is possible, with a fair degree of accuracy, to essentially "clean up" the Ramey memo and make it readable when the digital image is processed using optical character recognition computer software, as researchers did in this case.

In a July, 2000 letter to the editor of the *MUFON UFO Journal*, New Mexico State Director Dr. Donald Burleson commented on the ongoing optical character recognition and analysis of the Ramey memo, as well as relating an exchange with an audience member at a March, 2000 lecture held in Roswell.

When Dr. Burleson was speaking on the subject of the Ramey memo and the enigmatic "TEMPLE" signature line, he mentioned that, as of that time, the identity of the individual or individuals code named "TEMPLE" was unknown. It was at this point that Dr. Burleson received his answer directly from his audience, as he recounted in his correspondence with the *MUFON UFO Journal*. The following quote is attributed to the man, unnamed by Dr. Burleson in his letter, who spoke up following the March, 2000 lecture.

> *"I know who that is." When I asked him to explain,*
> *he said, "It's a code name for J. Edgar Hoover."*

When Burleson later interviewed this gentleman following his presentation, he learned that this individual had been attached to Strategic Air Command (SAC) headquarters in the 1960's, where he had seen numerous items of correspondence coming through the office with the TEMPLE signature line, and the witness was told at the time that the TEMPLE documents all originated from Hoover's office.[30]

In the interest of fairness, it should be noted that, to the knowledge of this writer, no evidence has yet been presented that either confirms or refutes the assertion that the code name TEMPLE was in fact a code name for FBI director J. Edgar Hoover, either at the time of the Roswell Incident or at any point in the future. It would seem that the TEMPLE connection to FBI director Hoover is inconclusive.

The passage of time, advancement in technology, and ingenuity of tenacious researchers have clearly made a compelling argument for the inclusion of the Roswell Incident alongside other forensic evidence cases.

The Roswell matter will seemingly always be the proverbial poster child for the field of ufology. Perhaps there will be a day in the future where evidence comes to light which clarifies the scientific document analysis done by Dr. Donald Burleson and others on the Ramey memo on July 8, 1947. It seems that only time will tell.

(Endnotes)

[30] Letter from Don Burleson to MUFON UFO Journal, July, 2000

6. The Trent Photos

On May 11, 1950 in McMinnville, Oregon, Paul Trent snapped what might be the single best known set of UFO photographs in history.

At approximately 7:30 PM, Trent's wife, Evelyn, was outside the home tending to a number of rabbits, which were kept on the property when she noticed an unusual object in the sky approaching from the north-northwest. Mrs. Trent had the presence of mind to call to her husband and ask him to bring their camera and photograph the object presently overhead. The object was in view long enough for Mr. Trent to snap two photographs.

The photos captured by Paul and Evelyn Trent that evening of are perhaps the two most important pieces of potential photographic evidence in the history of ufology. This argument can be made because the incident at McMinnville occurs so early in the modern UFO era just thirty-four months after the alleged crash of an unknown craft at Roswell, New Mexico, in 1947.

The trent photos as the standard for all other photo evidence cases

The Trent photos have become, over the span of fifty plus years, the standard for other UFO cases where photographic images are present. The McMinnville case was perhaps the first photographic evidence UFO case to be subjected to intense scientific scrutiny. The photos were published less than four weeks after the incident, when the editor of the local newspaper, the *McMinnville Telephone-Register*, stated,

"Expert photographers declared there has been no tampering with the negatives. The original photos were developed by a local firm. After careful consideration, there appears to be no possibility of hoax or hallucination connected with the pictures. Therefore, the Telephone-Register believes them authentic."

The Trent Photos have become the standard by which all other photo evidence UFO cases are judged, in the opinion of this writer, because they are, invariably, the most iconic images present in the literature, as well as having been captured early in the modern UFO era.

The Trent Photographs have withstood scientific scrutiny for over a half century. For this reason, the Trent Photos and the incident at McMinnville form, in the opinion of this writer, the bedrock of photographic analysis cases and should be considered the standard against which all other cases are measured.

The initial investigation and University of Colorado study

After being published in *The Portland Oregonian* (June 10, 1950), *The Los Angeles Examiner* (June 11, 1950), and *Life* Magazine (July 1950 issue), the Trents' original photos subsequently went missing from the files of International News Photo Service and later UPI, until they were located by the US Air Force sponsored University of Colorado UFO Study, otherwise known as the "Condon Committee", which operated from November, 1966 through November, 1968, headquartered on the campus of the University of Colorado in Boulder. It should be noted that this study has generated much controversy since the report was issued in 1968. The reader is invited to study the report for themselves. A fairly comprehensive overview of the Condon Report can be found at the popular internet encyclopedia Wikipedia.[1]

The Colorado study is commonly known in the field of ufology as "The Condon Committee", after nuclear physicist Edward Uhler Condon, who was, at the time, professor of physics at the school.[2]

The McMinnville case was one of a very select number reviewed by the Condon Committee during its term. Of some 15,000 documented cases as of November, 1966, the Condon Committee investigated barely 100, or approximately 1 ½ % of the total.

Condon's personal feelings towards the UFO subject was that they were *"bunk"*, believers were *"kooks"* and those who wrote books on the subject should be *"horsewhipped"*. Despite this, his committee diligently researched the Trent photographs.

In the official conclusion written for the University of Colorado study, case investigator William Hartmann writes:

> *"Certain physical evidence, specifically relative photographic densities of images in the photographs, suggests that the object was distant; if the object was truly distant, a hoax could be ruled out as beyond the capabilities of the photographer."*

It is important to note the testimony and official conclusions of investigator Hartmann, given his comments in support of an "unexplained" determination with respect to the Trent photographs.

In his official capacity with the University of Colorado UFO project, Hartmann is perhaps the first professional investigator to look at the photographic evidence from McMinnville, deliberate on that evidence, and return a verdict of "inconclusive" or "undetermined".

It is vitally important to view the Hartmann testimony and conclusions through the lens of the larger Condon Committee study, since Hartmann's conclusions (and later criticisms of Condon's objectivity) were echoed not only by fellow researchers on the panel, but also by the prestigious American Institute of Aeronautics and Astronautics, the professional society of the aerospace industry. In its rebuke of Condon, the AIAA wrote:

"The opposite conclusion could have been drawn from its content, namely that a phenomenon with such a high rate of unexplained cases should arouse sufficient scientific curiosity to continue its study." [3] It is also notable that the AIAA, in its statement, puts the "unexplained" figure at appoximately 30%.

To put this thirty percent figure into perspective, during the twenty-two years of Projects Sign/Grudge/Blue Book, the US Air Force collected some 12,618 case reports. Of that number, 701 cases were listed as "unidentified", leaving Project Blue Book with an "unsolved" rate of 18% as of December 17, 1969, when the project was officially terminated by the Secretary of the Air Force.

TOTAL UFO SIGHTINGS, 1947 - 1969

YEAR	TOTAL SIGHTINGS	UNIDENTIFIED
1947	122	12
1948	156	7
1949	186	22
1950	210	27
1951	169	22
1952	1,501	303
1953	509	42
1954	487	46
1955	545	24
1956	670	14
1957	1,006	14
1958	627	10
1959	390	12
1960	557	14
1961	591	13
1962	474	15
1963	399	14
1964	562	19
1965	887	16
1966	1,112	32
1967	937	19
1968	375	3
1969	146	1
TOTAL	12,618	701

Unidentified Cases, Project Blue Book

Courtesy The Black Vault

The question then remains:

Why would the United States Air Force close two official investigations into the UFO topic, where there were unsolved rates of 30% and 18%, respectively?

Perhaps there is no logical explanation.

The Maccabee Analysis

It was not until the early 1970s that the McMinnville case was re-examined by UFO researcher and physicist Bruce S. Maccabee.

Dr. Maccabee holds a doctorate degree in physics from the University of Maryland, and is employed as an optical physicist by the United States military.

Maccabee's decision to reopen the McMinnville case in late 1973 makes sense from an investigative standpoint for a number of reasons.

First, Maccabee's professional expertise, including his advanced training in physics, as well as his work in optics for the United States military, gave him a unique skill set which he could apply to his research. He could also apply this specialized knowledge to the McMinnville case in concert with advances in photography that had occurred in the nearly quarter-century between the original event and the beginning of his investigation.

It is of particular importance to note that multiple individuals involved in the Trent matter came to the opinion that the photographs were authentic and not the product of a hoax over a period of twenty years, long before Maccabee initiated his own investigation.

Bruce Maccabee's analysis has at its starting point the previously published conclusion of William Hartmann, the photo analyst with the Colorado UFO Study in 1968. As it was Hartmann's professional opinion that the photos were likely genuine, the scientifically responsible thing, in Maccabee's mind, was to either confirm or disprove Hartmann's analysis through an independent investigation of his own.

Maccabee's earliest analysis of the images in this case came in 1975, when he located the original Trent negatives, ironically with the incidental help of well-known debunker Phillip J. Klass. At some point between the conclusion of the Colorado Study in 1968 and Maccabee's analysis of the Trent negatives in 1975, the negatives had come into the possession of Phillip Bladine, the editor of the McMinnville newspaper.

There was some debate, given the shadows visible in the original prints, whether the photos might not have been taken in morning hours, rather than the evening, as the Trents had claimed.

In an overly simplified explanation, Maccabee's initial photo analysis was an updated "re-analysis" of the Hartmann conclusions, specifically centering on the image density of the Trent negatives.

The image density is tied directly to the exposure level of the film. Based on the conditions of the surrounding area at the time – an overcast evening – fits Maccabee's initial assertion that the *"gamma"* level (image density/image brightness of the negative) of the Trent negatives were lower than Hartmann's initial estimates, which were used in his calculations for the Colorado study.

Through his extensive knowledge of the photographic business, Dr. Maccabee was able to make a fairly educated guess as to the film likely used by the Trents at the time the photos were taken and, consequently, was able to make a fairly accurate estimation of the film's inherent properties.

In his report, *McMinnville: Scientific Analysis of the Most Famous UFO Photographs Ever Taken*, published by the Fund for UFO Research in 2000, Dr. Maccabee states:

"... Other possible film types are Plus-X and Plus-XX, both Kodak films, but the exposure curves of these are similar to that of Verichrome; measures of the fog density suggest that only Plus-XX and Verichrome are compatible with densities found in unexposed regions; Verichrome was the least expensive, hence most likely to be used; Verichrome has a low sensitivity to red light... Since the negatives are pale, that is, the density range starting from the fog level is not as large as expected for a sunlit day, I have assumed that the negatives were slightly underdeveloped and have, therefore, used an exposure curve for gamma = 0.6., even though it was standard procedure to develop to a gamma of about 1."

Trent photo #1, McMinnville, Oregon, 1950.
(Courtesy Black Vault Encyclopedia Project)

The Trent Photos in popular culture

The Trent Photos, as a result of surviving scientific scrutiny on multiple levels since 1950, have also become part of popular culture and have been featured repeatedly in both the literature common to the UFO research community as well as the books, magazine articles, and cable television specials that have become much more common in recent years.

Perhaps the case's most notable appearance in the mainstream media in recent years was its inclusion in an hour-long panel discussion debate on *Larry King Live* on July 6, 2005. The panel on the evening in question included noted UFO researchers John Schuessler, then MUFON International Director, Rob Swiatek and Dr. Bruce Maccabee representing the Fund for UFO Research, along with noted abduction researcher Budd Hopkins. Seth Shostak, director of SETI, The Search for Extraterrestrial Intelligence, was also on the panel, espousing what could be called a moderate skepticism. Additionally, Harvard psychologist Dr. Susan Clancy rounded out the panel, in what can only be described in the role of outright debunker of the UFO phenomenon.

The Trent photo case, in fact, provided what is perhaps the most telling moment of the hour-long program. Veteran researcher Robert Swiatek quite succinctly describes the case as follows:

> *"These photos have passed the muster over many years of study by many experts, including Dr. Maccabee here. Now the thing I wanted to qualify about UFO photographs is that in and of themselves they don't prove we're dealing with ETs. Any photograph can be hoaxed. The good thing about photographs like McMinnville is that all the obvious explanations for how the photographs might have been hoaxed or what these photographs show have been ruled out. i.e.: a model suspended from the wires. It's certainly not an unusual cloud. It's not something that the witnesses, threw in front of them, like a Frisbee, and photographed. Extensive research by private sector scientists like Dr. Maccabee and by the University of Colorado study back in the '60s, couldn't explain these sightings and they're representative of what people see."[4]*

The Trent Photo case after nearly sixty years

The photographs shot at McMinnville in June of 1950 are perhaps the gold standard in the field of ufology for cases with a photographic component due to the fact that they have been subjected to, and have withstood, scrutiny by the scientific community since the very beginning.

In the nearly six decades since Paul and Evelyn Trent captured a saucer-shaped craft on film, the photographs have withstood scientific scrutiny, on at least three separate occasions, by independent researchers with their own views of the phenomenon coming to their own distinct conclusions.

The initial investigation, conducted almost exclusively by the staff of the *McMinnville Telephone-Register* newspaper at the time of the event, is of vital importance in that this initial cursory investigation would pave the way for future inquiries.

The local investigation was conducted within weeks of the photos being taken. This is important because the incident was fresh in the minds of all involved. This is helpful because the investigators could proceed directly, rather than having to reconstruct the sequence of events after the fact.

It can be argued that it was inappropriate or perhaps even unethical for the local newspaper staff to undertake a cursory investigation of the photos, given that McMinnville was then, and remains today, a small town and it could have been seen as a conflict of interest. The investigators in this case had personal relationships with the Trent family. These were friends and neighbors – people who would pass one another in the street, wave hello, and, quite possibly, have any number of other collateral contacts throughout the course of their daily routines.

Despite these reasons, I believe an even more persuasive argument can be made in defense of the initial investigation conducted into the Trent Photographs by the staff of the *McMinnville Telephone-Register*. The Trents were local. This object, whatever it was, was photographed, quite literally, in their backyard. Not only was this unknown aircraft a matter of scientific and aviation interest, it was also a matter of local interest. It could have appeared over any other home in McMinnville, which made it a matter of local curiosity.

The second investigation into the McMinnville photographs, conducted during the University of Colorado's UFO Study, expanded on the initial research conducted in McMinnville, Oregon some sixteen years previously and set the stage for arguably the most comprehensive study of an individual case in the history of ufology.

The most recent investigation into the McMinnville case, conducted by Dr. Bruce Maccabee over a period of nearly twenty years, is not simply an in-depth study of a case from the annals of UFO research, but a methodical recreation of events, recitation of the facts, and twenty years of additional expert opinions.

While at times overly technical for the average person, Maccabee's personal investigation into McMinnville is exactly the treatment that the case deserves – all areas of the case are documented and every single criticism raised by previous investigators is answered expertly and in copious detail. The science behind the case is solid, even if previous investigators may have overlooked or ignored certain details.

Fifty eight years have passed since the event at McMinnville. Fifty eight years have passed since Paul Trent had the curiosity, quickness, and foresight to grab a small camera, race into his yard, and snap two hurried photos that would inspire curiosity, debate, and, ultimately, scientific inquiry, even years after his death.

Though not much on their face, the case of the Trent Photos has truly laid the groundwork for the forensic investigation of questioned images in the field of ufology, and will be, in the opinion of this writer, the standard-bearer for photographic evidence cases for years to come.

(Endnotes)

[1] http://en.wikipedia.org/wiki/Condon_Report

[2] http://en.wikipedia.org/wiki/Edward_Condon

[3] (Roesler, 2006)

[4] (Swiatek, 2005)

7. The Washington National Airport Sightings

Perhaps the most frequently asked question in the history of ufology is not whether the possibility exists that there is life elsewhere in the universe, but rather *"If we are being visited, why don't 'they' just land on the White House lawn?"*

In July, 1952, this highly unlikely scenario nearly came to pass.

While the events that transpired during the summer of 1952 in the skies over the District of Columbia were perhaps the best-known events of this period, the Washington National Airport sightings were anything but an isolated incident.

In his 1999 paper entitled *Acceptance of the Incredible: The 1952 Washington National Airport Sightings*, veteran UFO researcher Rob Swiatek details the growing trend of UFO sightings in the United States, beginning that January, up until the time of the National Airport events at the end of July.

As Swiatek notes, the US Air Force's official investigation into the UFO phenomenon, Project Grudge, was, at that time, underutilized and was drawing a very small number of official reports (around 10) per month for the entire United States. Events in the early part of the year would cause that number to increase by 100%. The restructuring of Project Grudge, as well as its re-branding as Project Blue Book, was a direct result of the increase in activity during the first quarter of the 1952 calendar year.

Even with the increase in case reports, Blue Book was egregiously understaffed in the early days of 1952. According to Swiatek, during the first quarter of 1952, Project Blue Book's official staff consisted of five individuals – two airmen, two civilians, and an official liaison officer stationed at the Pentagon.

At the helm of Blue Book at that time was Air Force Captain Edward Ruppelt, returning to active duty following service in the Second World War, during which time he earned several commendations, including five battle stars, two theatre combat ribbons, three air medals, and twice was awarded the Distinguished Flying Cross.

Ruppelt's military commendations are noteworthy, particularly because it is partly on these merits, one can assume, that the young Air Force captain, also one of the first military men trained in radar, was handpicked to head the re-branded Project Blue Book.

Captain Ruppelt is also key to the story of the Washington National Airport sightings, as the incidents were chiefly radar-oriented case reports.

On two successive weekends in July, the District of Columbia and the Washington metropolitan area were the setting for perhaps the most spectacular visual sightings, not only of their time, but in the history of the phenomenon.

Restricted airspace and military response

As any visitor to the Capital region will tell you, by virtue of its size, as well as its role as the seat of government, the entire District of Columbia is subject to some very unique, very necessary restrictions on commercial air travel.
What might be considered a "no-fly zone" by the general public is officially termed as a "Prohibited Area" by the Federal Aviation Administration.[35]

The FAA's Prohibited Areas in the Capital region are permanent due to Washington's status as the seat of government and, as such, frequent destination for foreign heads of state and other dignitaries. For matters of clarification, other noteworthy areas designated as "Prohibited" airspace in the Capital region include the National Mall, the United States Capital, the White House, and the U.S.

Naval Observatory, which serves as the official residence of the Vice President of the United States. Camp David, the presidential retreat located in Thurmont, Maryland, also has a permanent "Prohibited" designation.

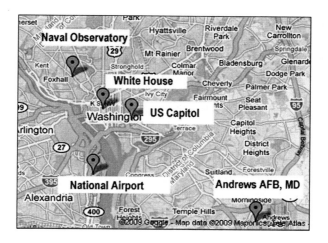

Map of the Washington area, showing many of the areas affected
by the July 1952 sightings.
Courtesy Google Maps (Mapped by the Author)

In addition to those areas classified as prohibited, the entire District of Columbia, as well as portions of the States of Virginia and Maryland are part of the FAA-designated Flight Restriction Zone surrounding Washington National Airport. This restriction encompasses all airspace within fifteen nautical miles of Washington National Airport, up to an altitude of 18,000 feet.

While perhaps overly technical, the distinction is nonetheless necessary, given that various sites adjacent to Washington National Airport are part of prohibited air space and that certain sites, specifically site P-56 (adjacent to the White House and U.S. Capitol Building), were compromised on several occasions during late evening/early morning hours of July 19-20 and July 26-27, 1952, in very real terms creating a national security incident in the skies above the Nation's Capital.

The national security implications of the UFO topic became apparent to the Blue Book staff, if it was not already aware, during the month of July, 1952, culminating in the events over National Airport on the weekends of July 19 and 26, respectively.

According to Richard Dolan's *UFOs and the National Security State,* there was what investigators term a "fly-by" sighting during the early morning hours of Sunday, July 13. A National Airlines pilot approximately sixty miles southwest of Washington reported a blue-white light approaching his aircraft, coming within two miles of the aircraft. Alarmed, the pilot radioed the Civil Aeronautics Authority, seeking advice. After a brief period, the unknown light *"took off up and away like a star at an estimated speed of 1000 mph".*

While not directly related to the events over Washington National Airport, the earlier reports from the capital region, as well as the State of Virginia, do have a bearing on the Washington National Airport cases.

The earlier Washington and Virginia cases show, if nothing else, that the amount of activity reported was escalating.

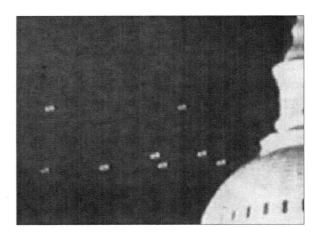

Unidentified objects buzz the US Capitol, 1952. Then, as today, Northern Virginia and the District of Columbia are part of a "Prohibited Area", as defined by the FAA.
Courtesy BJ Booth / UFOCasebook.com

The July 19-20, 1952 Events

The evening of Saturday, July 19, 1952 was the beginning of perhaps the single most fantastic and, for the lack of the more apt description, the most awe-inspiring UFO event on record.

While in some respects a photographic evidence case, the events over National Airport were primarily radar-visual sightings and as such, the story truly begins with the radar operations at Washington National Airport.

The radar operations at National Airport on that Saturday evening fell to air traffic controllers Ed Nugent and Harry Barnes, who began an eight hour shift in National's radar room at 11:00 PM on the evening of the 19th.

At exactly 11:40 PM, seven blips appeared on the long range radar scope at National Airport, originating in the southwest quadrant of the scope, in the area due east and slightly south of Andrews Air Force Base, Maryland, moving at a speed of approximately 100 miles an hour.

Under normal circumstances, this would not be cause for concern, or even appear out of the ordinary, especially for National Airport.

What made this series of blips not only highly unusual, but also a national security concern was the speed at which the blips changed course – within ten second sweeps of the scope, two of the unidentified objects accelerated suddenly, vanishing from view.

Given the distance between the points on the radar scope when the subject was seen, a staggering speed was determined – this object was calculated at a speed of approximately seven thousand miles per hour.

Confirming the unidentified blips on both of National Airport's radar terminals, Barnes, the senior air traffic controller at National Airport, notified the Air Defense Command and contacted Andrews, which also confirmed tracking the objects on its radar. In his conversation with Andrews personnel, Barnes asked if

Andrews would be sending military planes into the air to track the unidentified objects. The response from Andrews was negative. The Air Force was unable to scramble aircraft from Andrews, as the base's main runway was undergoing repairs at the time. As a result, the Air Force instead said that planes would be scrambled, but from New Castle Air Force Base outside of Wilmington, Delaware, rather than from Andrews.

Map showing National Airport, Andrews and New Castle Air Force Bases
Courtesy Google Maps. (Mapped by the Author)

The flight time between Wilmington, Delaware and the District of Columbia is approximately thirty to forty-five minutes. The aircraft en route from New Castle had orders to intercept any unknown aircraft that might be present in the airspace over Andrews or the District.

As Swiatek notes in *Acceptance of the Incredible*, and perhaps the most notable occurrence that evening, was the *"wanton disregard for the prohibited flight corridors scattered about the capital area"* shown by the objects tracked over both National Airport and Andrews Air Force Base. Of specific interest is the area of airspace designated as "P-56", which includes airspace directly above the US Capitol building and the White House.

At the time, Maj. Donald E. Keyhoe, perhaps best known for founding NICAP, wrote that two objects were seen over the White House, with a third unknown being seen near the US Capitol.

While the dramatic appearance of anomalous objects over perhaps the two most important buildings in the nation's capital make this case particularly memorable, it is important that researchers of the phenomenon not forget that the events over Washington, DC can be classified in any number of categories.

On the one hand, they are clearly radar related sightings. The radar component of these cases is vitally important from the standpoint of documentation physical evidence. Of equal importance, however, are the visual sighting reports which came in to air traffic control staff at National Airport over a period of hours from the late evening of July 19, 1952 to the early morning hours of the following day.

A Capital Airlines flight, departing from Washington National Airport at 1:00 AM on the morning of Sunday, July 20, 1952, approximately one hour and twenty minutes after the first radar sightings, reported a number of objects that were clearly visible in close proximity to the aircraft. Capital Airlines pilot S.C. Pierman reported several objects that darted and flashed in the immediate vicinity of the aircraft. He radioed the movements of these objects down to the radar room at National Airport, as later testified to by chief controller Harry Barnes.

These same objects were, apparently, reported approximately an hour later by personnel in the vicinity of Andrews Air Force Base in Maryland.

Ronald Reagan Washington National Airport at night, 2005.
Courtesy Wikimedia Foundation

Staff Sgt. Charles Davenport, as noted by researcher Robert Swiatek, later testified as follows:

> *"On the morning of the 20th of July at approximately 0200 I saw a strange light south of Andrews AFB traveling from east to west at a terrific rate of speed. ... At times it would appear to stand still then make an abrupt change of direction and altitude. I was unable to tell what it was so I called Andrews tower and asked if they had spotted or knew what it was. They finally saw it for a few seconds off runway 28 about that time it shot out of sight at terrific speed. It's [sic] color was an orange-red."*

Perhaps the most important factor in this case is the visual sighting reports which came in from aircraft in the air, augmenting what was being reported by radar operators on the ground at multiple locations. It is also important to understand that the initial series of reports over Washington took place over a period of several hours, rather than being confined to a smaller time span.

From start to finish, the anomalous lights reported over Washington's airspace were seen for approximately six hours, from 11:40 PM until approximately 5:30 the next morning.

The July 26-27, 1952 Events

Exactly one week after the flyover of July 20, strange lights and objects were again seen over the District of Columbia.

Then, as before, the objects were reportedly seen in the airspace of National Airport and continued on a flight path that included many Washington, DC landmarks.

The initial report was made at approximately 8:15 in the evening by the pilot of a National Airlines flight in the air in the vicinity of the District. Both the pilot and stewardess onboard reported seeing several objects resembling glowing cigarette embers high above them, moving at an estimated speed of 100 miles an hour.

Within a short period of time, both Andrews Air Force Base and Washington National Airport personnel would independently confirm as many as a dozen objects matching the description given by the National Airlines flight crew.

While the objects reported moved at random and at a lower speed then aircraft of the period, they did, nonetheless, exhibit characteristics which led training professionals with experience in aviation to draw the conclusion that said objects were operating under some level of intelligent control.

At 9:46 PM Eastern Time that evening, a Civil Aeronautics Authority pilot reported five objects which matched the description given by the National Airlines pilot.

Washington Post, July 28, 1952.
Courtesy Black Vault Encyclopedia Project

For a period five hours, between 10 PM July 26, 1952 and 3 AM on the morning of July 27, 1952, these objects were tracked on radar from several locations, including Washington National Airport and Andrews Air Force Base, Maryland.

As in the July 19-20 incident, the U.S. Air Force was alerted, scrambling aircraft from New Castle Air Force Base, Delaware, under orders to intercept any of the unidentified aircraft.

According to published reports, when the Air Force interceptors would close on an object, it would subsequently disappear from both the pilot's field of view and the radar scopes on the ground.

As with the July 19 incident, the F-94 pilots would run low on fuel, returning to Delaware.

Perhaps the most curious incident in the entire series of events occurred not in the air, but on the ground. Barely forty-eight hours after the latest incursion over Washington's skies, the Air Force held a large scale press conference, which included remarks from US Air Force Major General John Samford, director of intelligence, Captain Ruppelt of Project Blue Book, and Major General Roger Ramey, perhaps best known within the field of ufology for his role in the aftermath of the Roswell Incident.

While the press conference convened on July 29, 1952 claimed to offer the public and media answers, the panel, while in the unenviable position of facing the media to explain away a security breach of the highest order, left the general public with seemingly nothing in the way of answers.

The July 2002 incident

In what is perhaps the best and most succinct description of the Washington National Airport events, historian Richard Dolan, author of *UFOs and the National Security State*, discusses the 2002 flyover of Washington during a press conference called by the Paradigm Research Group at the National Press Club on April 25, 2005. Speaking of Washington events, Dolan stated,

> *"In July of 2002, right here in the Washington DC area, there was, yet again, an airspace jet chase by two F-16 jet interceptors which were chasing a bluish object right outside DC. This received some play in the Washington Post and a little bit of mainstream media, then, of course, fell into the black hole of the media, never to be discussed again. What were these objects that were able to out perform F-16 interceptors in a post-9/11 America? What is going on?"*

Dolan makes a fair point in his assessment of this incident, in that it occurred not one year removed from the September 11, 2001 terrorist attacks carried out over Washington, DC and New York City.

While the mention of the 9/11 terrorist attacks on Washington and New York may have caused some to blanch, it was certainly germane to the question Dolan raised at the National Press Club. In an age where national security was of the utmost concern, what technology could not only invade, for want of a better word, the most secure military airspace in the United States, but also outperform some of the best military aircraft in the United States arsenal?

When this additional information is taken into account, the July 26, 2002 aerial incident, which occurred over the District of Columbia and Northern Virginia fifty years to the day after the original Washington National Airport events, takes on added significance when viewed in conjunction with earlier events.

The National Airport events of 1952 alone are enough to raise concern, as they did at the time, considering that the unknown objects were tracked on radar from no fewer than three locations in the vicinity of Washington DC.

Dolan makes several valid points in his April 25, 2005 comments with regard to the 1952 and 2002 Washington UFO events, not the least of which being a very simple reference to the coverage received in the pages of the *Washington Post*.

In his piece entitled 'Alien Armada', published on Sunday, July 21, 2002, Post staff writer Peter Carlson detailed the 1952 events fairly well, especially considering the way in which the UFO topic is generally handled by the mainstream media.

Rather than making light of the UFO subject, Carlson and the Post ran a piece that treated the 1952 events as a matter of local history.

The fact that a newspaper such as the *Washington Post* would devote column space to a news item dealing with UFOs speaks directly to the fact that both incidents have, at their core, hard science.

The 2002 UFO incident over Washington paralleled the July 26, 1952 sightings in many respects, including the fact that the objects reported seemed to follow a nearly identical flight path, as well as, allegedly, flying in a formation similar to the 1952 reports.

The 1952 National Airport case in historical context

The 1952 National Airport sighting wave is a clear example of a reported UFO event that takes on new significance with the passage of time.

While this can be said for virtually any case in the field of ufology, it is particularly relevant with regard to the Washington, DC events.

Very few UFO cases can be traced back to the absolute exact location described in a case report. For one reason or another, memories fade, landscapes change, and urban sprawl encroaches on what was once wide open space.
This is clearly not the case with the 1952 Washington National Airport events.

Though renovated[36] in 1997, National's Terminal A remains from the time of the building's original construction, making it – in a forensic sense – evidence preserved from the time of the event.

Ronald Reagan Washington National Airport, 2007
Courtesy of the Author

During the research and writing process for this book, this writer flew on several occasions between Milwaukee, Wisconsin and Ronald Reagan Washington National Airport.

By happenstance, all of my flights into National Airport arrived at the A concourse, in the oldest wing of the airport. I couldn't help but wonder if my incoming flight had just taken the same flight path as those numerous unidentified targets nearly sixty years previously.

Personally speaking, the events of 1952 were constantly in mind with every arrival to and departure from the District. In a very real sense, I was on-scene at a cold crime scene, despite it also being one of the busiest airports in the country.

This is particularly interesting due to the fact that, as Reagan National Airport is situated between both the Commonwealth of Virginia and the District of Columbia, there is little, if any, room for future expansion.

Washington National Airport will not only continue to serve the residents of Virginia and the Nation's Capital as a vital piece of infrastructure, but may very well stand as a continuing reminder of the very curious events of the summer of 1952.

(Endnotes)

[35] http://www.sua.faa.gov

[36] http://www.metwashairports.com/reagan/about_reagan_national/history_3

8. The Heflin Photos

On the afternoon of August 3, 1965, California highway maintenance engineer Rex Heflin captured what has become, arguably, some of the best photographic evidence in the history of ufology.

During the course of his workday, Heflin would routinely drive the highway, documenting any repairs that his department needed to make, such as items obstructing the roadway which needed to be cleared, damaged signs or mile markers to be replaced, or areas in the roadway which needed repainting.

On that afternoon, Heflin's work detail took him to Santa Ana, California, located in Orange County, some forty miles south of the city of Los Angeles.

At approximately 12:30 PM that afternoon, Heflin pulled his work vehicle out of traffic to photograph a damaged road sign, as his earlier attempt to report the damage by radio was unsuccessful. As Heflin picked up his work issued Polaroid camera to photograph the damaged sign, he noticed a sudden movement in the sky to his left, on the periphery of his field of vision. Over the next two minutes, Heflin would capture four photographs that would eventually become some of the most widely circulated in the world.

It is the opinion of this writer that the fame garnered by the Heflin photos – both at the time of the event and at present – is a result of a number of factors taken in concert, a "perfect storm", so to speak.

The case of Rex Heflin, it can be argued, was not only a watershed in the field of ufology with regard to photographic evidence cases, but it is also of fundamental importance as a case study. Many hallmarks commonly seen in UFO cases across the globe appear, in one form or the other, in the Heflin case.

Principally, the Heflin case is a photographic evidence case. The photographs are clear and compelling evidence of something anomalous in the skies over Santa Ana, California in August of 1965.

It is important to note that while certain segments of the media may use the term 'UFO' as a catchall term to imply "extraterrestrial spacecraft", the research community does not take this view. The research community prefers to use the term "unidentified aerial phenomena" (UAP) or a redefinition of the term "UFO", with the letter U now denoting "unconventional".

This distinction is especially important with regard to the Heflin case, as Mr. Heflin never suggested that the object he captured on film was anything other than an experimental or top secret aircraft stationed at the nearby El Toro Marine Corps Air Station.

Marine Corps Air Station El Toro, 1993
Courtesy United States Marine Corps

It is fundamentally important to note that, at time of the initial investigation, Rex Heflin never suggested that what he had photographed was a spacecraft or in any way extraterrestrial.

While Heflin's opinion did change over time, it is important to understand that his opinion changed as a result not of fame or media attention, but due to the scientific and professional opinions of the men and women involved in researching his experience.

In order to properly understand the investigation that was conducted into the Heflin photographs, it is incumbent upon us to "begin at the beginning".

In this case, this means to first understand not just the photographs which are at the center of the Heflin case, but, first and foremost, the camera which shot those four images.

Understanding the capabilities of Heflin's camera is of the utmost importance when discussing the photographs themselves. Of course, by understanding the camera, one gains greater insight into the photographs which the camera created. Having a working knowledge of the make and model of camera allows you to understand exactly the conditions under which the photograph was created, thereby making the photographs themselves pieces of scientific evidence.

The camera in question was a Polaroid model 101, loaded with ASA 3000 film. For those unfamiliar with photography, the "ASA number" (later replaced by an ISO rating number) denotes a film's sensitivity to light. For example, a 35mm film with an ASA/ISO rating of 100 would be suitable, almost exclusively, for outdoor photography or photography taken under very bright conditions.

In very simple terms, the rating number assigned to a particular film is loosely analogous to rating a digital image by pixel size. Simply put, the higher a film's rating, the higher the image resolution.

Rather than discussing the complex mathematics involved in determining film speed and, given the fact that, as of February, 2008, Polaroid was no longer manufacturing or distributing instant film similar to the type used by Heflin, I will instead describe the properties of a high-speed film readily available on the market today with properties virtually identical to the ASA 3000 film used to take the photographs in question. While there are subtle differences between the 1965 Polaroid film which captured the questioned photographs and the modern black-and-white film, any such differences are negligible at best, allowing for a very accurate approximation.

Given that Polaroid Corporation no longer produces a film with an ASA/ISO rating of 3000, the closest equivalent on the market today seems to be TMax P3200, a 35mm black-and-white film produced by Kodak. A professional grade film, Kodak TMax P3200 has a slightly higher rating than the questioned film, but is analogous for purposes of demonstration.

According to Kodak's own documentation[37], the TMax P3200 film is designed to be used as a multi-speed film. This film, by design, was intended to be used for a wide range of applications, up to and including harsh weather conditions.

Given this versatility, it isn't surprising that a similar film would have found use under the various conditions that Rex Heflin and his coworkers would have encountered on a daily basis.

While subtle differences between the questioned film and modern film undoubtedly exist, there is enough similarity on a chemical level and with regard to performance that readily available film would suffice as an analog for the questioned film, which has long since gone out of production.

The initial investigation

The original investigation into the Heflin sighting is notable, perhaps, for its level of technical sophistication for the time, as well as the fact that it was kept as an open investigation long after various factions within the UFO research community of the 1960s and 70s, as well as a percentage of the mainstream media had branded it a hoax. What makes the Heflin case of particular importance is the fact that it has been researched more thoroughly than seemingly any other photo evidence case on record.

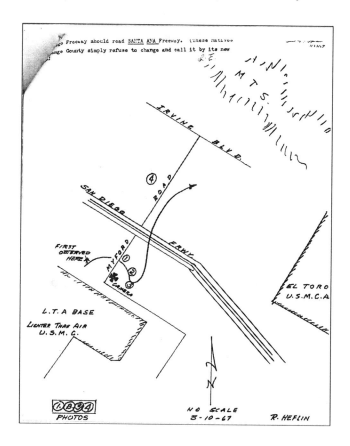

**Map showing the locations of the Heflin Photos, as it appeared
in Heflin's original report to NICAP.**
*Courtesy J.E. McDonald "Heflin" file, University of Arizona Library,
Special Collections Section*

The investigation into the Heflin photographs was conducted, principally, by researcher John R. Gray, Ann Druffel, and Idabel Epperson of the Los Angeles NICAP Subcommittee (LANS), with additional research being conducted by Drs. James McDonald and William Hartmann of the University of Arizona.

It is fair to say that the Heflin photo case is certainly one of the *"best evidence"* cases in the history of ufology, however, is far from the ideal. As a case which could be presented to a jury, the Heflin matter is potentially problematic for ufologists who would argue the case in the proverbial court of public opinion.

Perhaps the most apparent hurdle faced by researchers is the fact that even before investigators from NICAP became involved, the original photos had left Rex Heflin's possession several times, with several sets of first-generation copies being made.

Between the date of the event, August 3, and the beginning of the official NICAP/LANS investigation on September 20, no fewer than three sets of duplicate prints were made, as well as six sets of negatives from the original prints. In addition, Heflin also allowed a coworker to submit the original photographs to the Los Angeles office of *Life Magazine*, which, in turn, forwarded the photos on to New York City.

This series of events, in the opinion of this writer, did significant damage to the case, from a chain of custody prospective, in that the photographs left Heflin's possession for an extended period, with only the slightest documentation regarding destination, storage conditions, and other factors that could, from an evidentiary point of view, have significant bearing on the case from an investigative standpoint.

Chain of custody issues aside, the numerous sets of duplicate prints circulating at the time of the initial investigation would come to play a very valuable role in the investigation, although it was not immediately apparent. The multiple sets of duplicate photographs come into play precisely because of variation to the printing processes employed by the various individuals who retained Heflin's original prints prior to the beginning of the official NICAP investigation.

According to the official case file, amateur photographer Wayne Thornhill was apparently the second individual to make copies of the original prints. The Thornhill copies are of particular importance because, while of poor quality, the prints showed the entire image and were not cropped during processing. This is a very subtle, yet important distinction, as additional prints made from the Heflin originals, for publication in the *Santa Ana Register* were cropped, thereby altering the image ever so slightly. This is noteworthy because investigators would NICAP were able to verify the exact time that the first photograph was taken, using the uncropped photos as well as weather data from the Griffith Park Observatory in Los Angeles, among other sources.

During the initial investigation conducted by NICAP and its Los Angeles subcommittee, even more copies were produced, this time for researchers John Gray, Idabel Epperson, and University of Arizona physicist Dr. James McDonald. Of these copies, it important to note that differences between prints do exist, particularly between the Epperson and McDonald copies.

While these minor variations in print quality seem to have been very significant in the mind of Dr. McDonald, there are numerous explanations for the discrepancies between the Epperson and McDonald prints. The most likely explanation – in the opinion of this writer – for the dark, somewhat cloudy appearance of the McDonald images, compared with Epperson's copies, would seem to be outdated or overused developing chemicals. The apparent differences in appearance between the two most prominent sets of duplicate originals used in the investigation were the first of several issues that would complicate matters moving forward.

Concurrent with the NICAP investigation, Heflin's photographs came to the attention of the United States Air Force, in the person of Capt. Charles F. Reichmuth, who officially investigated Heflin and his sighting report for the US military. Reichmuth's interview of Heflin occurred on the afternoon of September 23, only days into the NICAP investigation. On the same day, Maj. Hector Quintanilla, head of Project Blue Book, issued a statement regarding the Heflin photos, saying, in essence, that Heflin's original statements did not match the calculations of Blue Book photo analysts, although Quintanilla conceded that measurements made by his staff were made using wire service photographs and thus, *"it was impossible to be absolutely correct in the analysis"*.

In the opinion of this writer, it is curious at best that the head of the U.S. Air Force's official UFO investigation would rely on wire service photographs as a source of information. It is equally interesting that Project Blue Book would disregard the official report of Capt. Reichmuth, who stated in his official report that he could find no evidence that would contradict character witnesses, who testified that Heflin was mature, professional, and trustworthy. Reichmuth went on to state *"from all appearances, he [Heflin] is not attempting to perpetrate a hoax."*

Within Dr. James McDonald's personal file, several documents exist detailing an ongoing dialogue with personnel at the Polaroid Corporation regarding various technical issues, including a somewhat contentious point of fact: Whether or not there was ever a period of time where the film packs produced for the model 101 camera did not contain reference numbers. Many of the scientists involved with the investigation were insistent that all of Polaroid's film packs on the market at that time contained pre-printed reference numbers. From the point of view of these researchers, said reference numbers would be the simplest way of determining which photos were shot with which film and in what sequence. In essence, what was being done was that a timeline of events was being established, just as would be presented in any court proceeding. Using this established timeline, researchers would then be able to test the reliability of Mr. Heflin's statement.

This is not to say that Heflin gave researchers any cause to feel that he was being anything less than truthful. Independent interviews conducted with friends, relatives, coworkers, and superiors all indicated that Rex Heflin was hard-working, conscientious, and would certainly not be the type of person to seek attention.

In a letter to Dr. McDonald dated February 28, 1968, a Polaroid employee was able to verify that there was, in fact, a period of time where film for the Model 101 camera was produced that lacked sequence numbers found on all other film packs in the possession of the Orange County Highway Department, Heflin's employer.

The Polaroid letter indicates that based on the serial number of the film, it had a date of manufacture of November, 1964, making the film approximately nine months old at the time the photos were taken.

While it is true that age affects the quality of certain film, there is nothing in McDonald's file to suggest that faulty film was over considered as a possible source of the Heflin images. The Polaroid letter of February 28 goes on to state that *"Mr. Heflin's report is entirely feasible."*

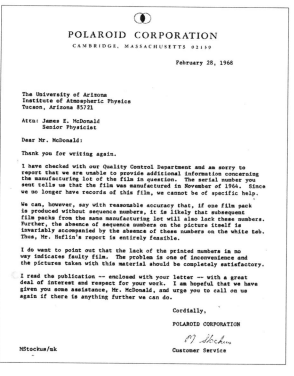

February 28, 1968 letter from Polaroid Corp. to Dr. James E. McDonald
Courtesy J.E. McDonald "Heflin" file, University of Arizona Library, Special Collections Section

Heflin and the men in black

Perhaps the most curious element of the Heflin photo case is Rex Heflin's encounter with seemingly anonymous government agents – figures known today as the Men In Black.

Contrary to the image portrayed in the 1996 film, the Men In Black, as they appear in other UFO case reports, often serve as agents of disinformation and intimidation, while the UFO witness, already operating outside their comfort zone, is left questioning both their own senses and past experiences.

Shortly after the initial investigation was convened by NICAP and LANS, Heflin was contacted by an individual claiming to be a colonel with NORAD – the North American Aerospace Defense Command – asking to meet with them and discuss the photos that he had taken.

As he had previously discussed the photographs with both the Air Force and US Marine Corps, Heflin presumably thought nothing of speaking with other government officials regarding his experience.

On the evening of September 22, 1965, two men in civilian clothes arrived at Heflin's home, presenting identification that, to Heflin, resembled the ID carried by US Marines at El Toro, except that it lacked a photograph.

There is no record, so far as this author is aware, of Heflin's meeting with the supposed NORAD officials, but this much is known: Heflin recounted the events of August 3, answered additional questions, and ultimately turned over photographs 1 through 3 to the men, with the expectation that the photos would be returned upon completion of their inquiry, as had been the case with both the Air Force and Marine Corps personnel.

The original prints were never returned, their whereabouts unknown until 1993, when they would reappear in Heflin's mailbox, in a plain envelope bearing no postmark.

With regard to the reemergence of the original photographs, perhaps the most interesting detail, in my opinion, is the condition of the photos upon their return.

By Rex Heflin's own admission, the photos were returned in near perfect condition, save for the word "ORIGINAL", which had been written across the tops of each photo in what appeared to be white grease pencil.

Given these facts, a few educated inferences can be made.

Firstly, the photographs were obtained from Mr. Heflin by individuals unknown through false pretenses.

Secondly, the photographs were apparently the subject of what was presumably an extensive study carried out by the person or organization who had taken custody of the prints twenty-eight years previously.

Lastly, it can be assumed that the photographs were returned in 1993 because they were no longer of value to the party or parties that had held them since 1965.

Following the return of the photographs, they remained at the Heflin residence in Northern California until they were entrusted to veteran ufologist Ann Druffel, who had been part of the original NICAP investigative team in 1965. Druffel retains custody of the photographs to this day.

Reanalysis of the Heflin Photos

With the original photographs in the hands of researchers for the first time in over thirty years, reanalysis of the images was now possible, using technology that could barely have been imagined at the time of the original NICAP investigation.

The modern investigative team consisted of veteran UFO researcher Ann Druffel, aerospace engineer Dr. Robert M. Wood, and Dr. Eric Kelson, a science professor at California State University at Northridge.

This team was uniquely suited to undertake a reanalysis of the original photographs for several reasons, not the least of which being, in the case of Ann Druffel, a firsthand knowledge of the case, which would provide the investigative team with an invaluable perspective on the case.

Aerospace engineer Robert Wood, it should be noted, was also involved in the original Heflin investigation to a degree, as McDonald's case file also mentions him as having been present at

LANS meetings of the period, so he, like Ann Druffel, was able to bring his own personal insights to bear when re-evaluating the evidence.

Additionally, the investigative team was able to trade on Wood's engineering background with Douglas Aircraft (later McDonnell-Douglas). Wood's technical expertise, in my view, is invaluable, given that his professional credentials include research and development work on both military and civilian aircraft. This level of experience would allow Dr. Wood, in essence, to serve as an expert witness, having also conducted an independent analysis of the questioned photos in the 1960's.

The final member of the re-evaluation team was Cal State-Northridge professor Dr. Eric Kelson, who, like Dr. Robert Wood, brought a technical background to the project. Dr. Kelson, however, brought a very specific and vital skill set to the project – his expertise with computers.

Kelson's expertise took on an added importance, given that Heflin's original Polaroids were again available, rather than second, third, or fourth generation reprints, which, in the past, were the only available prints from which to work.

While at times overly technical for the average person, the end result of the Druffel-Wood-Kelson analysis confirms Rex Heflin's account as being consistent with the photographs, and dispels the notion that Heflin had perpetrated a hoax.

Factors that rule out a hoax hypothesis

Several factors, in the opinion of this writer, rule out the possibility that Heflin was involved in perpetrating a relatively sophisticated hoax in this case.

The primary factor that would seem to suggest that the Heflin case is a genuine anomalous event rather than a deliberate hoax is the fact that the camera equipment used by Heflin would make a hoax exceedingly unlikely, especially considering that the means by which a hoax would be perpetrated were far more rudimentary in the mid-1960s compared to the technology available today.

Heflin Photo # 1. Perhaps the most iconic photograph in the history of Ufology. *Courtesy Fund for UFO Research, Inc.*

This is, of course, presuming that Rex Heflin would be the type of individual to perpetrate a hoax. From all accounts, including several independent investigations done and countless interviews conducted at the time of the initial event, all parties concerned came to the conclusion that while Heflin had "an offbeat sense of humor"[38], he would certainly not jeopardize his employment to perpetrate a hoax.

Is it possible to re-create (or closely approximate) the Heflin photographs using digital technology available today? Certainly. The question is not whether questioned photographs some forty-plus years old can be re-created using modern technology, but if those same photographs could have been hoaxed using technology and materials readily available at the time of the event.

While there were a number of individuals, most notably Dr. William Hartmann of the University of Arizona, who questioned the authenticity of the photographs, the objections raised have since been refuted through technological advancement and further study.

As detailed in 2000 by researchers Ann Druffel, Dr. Robert Wood, and Dr. Eric Kelson, Hartmann attempted to replicate the Heflin photos using a similar camera and a model hung on a string, as well as suspending the lens cap from a Leica camera from a thread. While both of Hartmann's attempts were similar in clarity to the Heflin photo, the authors note that the string suspending the model was plainly visible. In addition, the "wedge of light" seen in the second Heflin photo could not be reproduced.

Given that Hartmann could not reproduce accurately all aspects of Heflin Photo # 2, in the opinion of this writer, Hartmann's official conclusion, as presented to the University of Colorado UFO Study (The Condon Committee), is invalidated. As noted by Druffel, Wood, and Kelson, Hartmann's conclusion, which claimed that his own results closely matched Heflin's second photo was a misstatement at best, as well as *"an early example of the naïve assumption that just because a photograph could be ostensibly faked, that it was faked"*.

It is this writer's opinion that both Dr. Hartmann and the Colorado UFO study would have been better served by simply noting their dissension from researchers Gray and McDonald, among others, and recommending that judgment be reserved pending further study. Had Hartmann and the Colorado study taken this option, perhaps the entire project would be viewed in a more positive light today.

Furthermore, Druffel, Wood, and Kelson meticulously detail all aspects of the Heflin case, including details about the photographs which only have become apparent thanks to advancements in imaging technology.

Specifically, when one takes into account both the original case file, meticulously assembled by the late Dr. McDonald, as well as the reanalysis performed at the beginning of this decade, it is this writer's opinion that the conclusions reached by Druffel, Wood, and Kelson – primarily that the photographs are entirely consistent with the original witness narrative – are valid.

Additionally, it may seem like an overly simplistic explanation, but nowhere in the case file does it suggest that Heflin possessed anything beyond a working knowledge of camera equipment, let alone enough knowledge to operate darkroom equipment with the skill and precision necessary to composite a hoaxed photograph.

Heflin Photo # 2
Courtesy Fund for UFO Research, Inc.

The Heflin Legacy

The Heflin UFO case, in the opinion of this writer, is perhaps the single greatest example of science put to use to examine anomalous phenomena. The most interesting piece of anecdotal evidence, in my view, is most definitely the case file compiled by Dr. James McDonald, consisting of over 250 typed pages, including the initial NICAP materials, as well as material from the files of Project Blue Book.

It is important to understand that the Heflin case is much more than an isolated incident. In my view, this case is very much a "textbook" example of both investigative techniques seen in later cases, as well as an example of certain anomalies which have presented themselves in other cases as well.

For example, both McDonald's case file as well as the photographic reanalysis published in 2000 by Druffel, Wood, and Kelson make mention of an unrelated photographic evidence case, reported in Fort Belvoir, Virginia, which also displayed some of the same anomalous features, including an elliptical smoke ring.

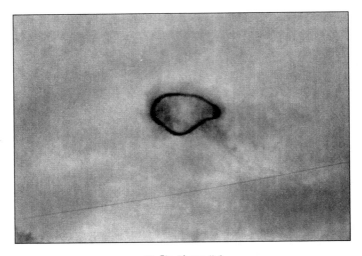

Heflin Photo # 4
Courtesy J.E. McDonald "Heflin" file, University of Arizona Library, Special Collections Section

While there is nothing concrete to suggest a connection between the two cases, a comparison of Heflin Photo #4 and a photograph from Fort Belvoir show interesting similarities.

Case studies aside, the Rex Heflin photo case occupies an important place in the modern history of ufology in that it can be seen, in many ways, as being a bridge between certain *"first generation"* cases, which involve very basic photographic images and later events, which encompass any number of photographic techniques, including 35mm print photography, as well as the digital photography, radar-visual, and video evidence cases that would follow.

Despite advancements in technology over the intervening four decades since the event, the Heflin photo case is still as relevant today to the field of ufology as it was at the time of the event.

Fort Belvoir, Virginia (1957)
Courtesy UFO Casebook.com / BJ Booth

In a very real sense, the scientific research undertaken by John Gray, Dr. James McDonald, the LANS investigators, and others has helped to shape the debate surrounding photographic evidence UFO cases moving forward to the present day. This case, in the opinion of this writer, is perhaps the single best example of a scientific investigation involving questioned photographs to date.

(Endnotes)

[37] http://www.kodak.com/global/en/professional/support/techPubs/f4016/f4016.pdf

[38] (Druffel, Wood, Kelson, 2000)

9. The Michigan Sightings

A single UFO sighting event is note worthy in any locale in any nation across the world. These events take on added meaning when they are seen repeatedly in a small geographic area in a short span of time.

In the terminology of the UFO research field, repeated or mass sightings such as this would be termed "*a wave*".

Such a wave occurred during the spring of 1966, when UFOs were reported over virtually the whole of lower Michigan, from Grand Rapids in the west to the Detroit metro area in the east. However, in this instance, the cases of interest come from points in between, the area surrounding Ann Arbor, Michigan, which saw numerous UFO events reported in a one-week span from March 14 through March 20, 1966.

What makes these cases particularly interesting, in the opinion of this writer, is the fact that both the Milan and Dexter, Michigan cases involved trained observers, who would meet the burden of proof for an expert witness, as defined by the legal system. While it is purely supposition at this point in time, the fact of the matter remains that when an unexplained event, such as a UFO sighting occurs, the testimony of law enforcement and military personnel would hold greater weight, at least from a technical standpoint, than an average citizen witness.

The Michigan sightings of 1966 are important pieces of the overall puzzle on a number of fronts.

Chiefly, these cases are of historical value, given that a photographic record exists, which may be used as a point of reference or comparison to other photographic cases.

Secondly, the expert witnesses associated with these cases, both law enforcement and military, laid the groundwork for subsequent investigations, specifically by astronomer J. Allen Hynek, then the scientific consultant to Project Blue Book.

As evidenced in later cases, professionally trained observers, such as law enforcement officers, are valuable witnesses due to finely honed observation methods and detailed writing skills necessary in the course of their daily professional lives. In this respect, both the Milan and Dexter, Michigan events excellent case studies, as witnesses to both events were law enforcement officers and hence, trained observers. This alone makes the claims of Dr. Hynek and Blue Book seem downright laughable in retrospect, further underscoring the point that the Air Force was seemingly going out of its way to explain UFO events, no matter how flimsy that explanation would appear on further examination.

Lastly, these events are of historical interest, given that the frequent press coverage caught the eye of then-Congressman Gerald Ford of Michigan's fifth Congressional District. It is due to Ford's involvement that a copious amount of material remains preserved for public consumption, thanks to his status (at the time of the event) as House Minority Leader.

This interesting confluence of circumstances, in my view, makes the 1966 Michigan sighting wave a unique, intriguing, and certainly valid case study.

17 March 1966: Milan, Michigan

On the evening of March 17, 1966, one of the more intriguing pieces of photographic evidence was captured when Washtenaw County Deputy Sheriff David Fitzpatrick captured what appear to be several curious streaks of light in the night sky in Milan, Michigan.

The Milan, Michigan UFO case is of particular interest due to the fact that the photographer and primary witness was a local law enforcement officer. Rightly or wrongly, law enforcement officials such as Fitzpatrick are often viewed by the general public is being more credible than a citizen witness and are subsequently held to a higher standard, due to their careers, which require that they be trained observers.

Regardless of one's opinion as to the reliability of eyewitness testimony, as there is quite a bit of evidence within both the fields of law enforcement and psychology that would cast doubt on eyewitness testimony alone, it is important to remember that, in this case, Deputy Fitzpatrick was a member of the local sheriff's department and, therefore, likely intimately familiar with the area, which would make his testimony quite valuable.

The interesting fact in the Milan, Michigan case, in my view, is the fact that it not only involves a questioned photograph, but also the fact that perhaps the most famous reference to this case stems from a press conference given by Dr. J. Allen Hynek, in which he dismissed the witness reports and the sighting as a whole as a misidentification of the planet Venus.

I say this because it is perhaps the best example of what, to this writer, was common practice for the staff of Project Blue Book – clearing case reports from the ranks of the unidentified or unexplained by relying on public statements issued by Dr. Hynek.

Lastly, the Milan, Michigan events are incredibly significant, though often overlooked, as they indirectly gave the field of ufology perhaps its single greatest champion in Hynek, who became convinced, in large part due to his experiences with the Michigan events, that the UFO subject was much more than a series of misidentifications and was, instead, a subject worthy of scientific scrutiny.

20 March 1966: Dexter, Michigan

Not three days after the Milan UFO event made headlines, a UFO event would be reported in the nearby town of Dexter, Michigan, some twenty-six miles away.

The Dexter, Michigan UFO event is noteworthy for several reasons. Chief among these is the fact that both the Milan and Dexter cases involved not only local law enforcement, but, in fact, the same deputy sheriff, David Fitzpatrick.

As mentioned previously, the towns of Milan and Dexter, Michigan are separated by only a very short distance, which would make it quite conceivable that a single deputy sheriff – in this case, Deputy Fitzpatrick – could be involved in multiple calls stemming from reports of strange objects seen over the local area.

The Dexter sighting is perhaps best known in the field of ufology for being the genesis of the statement attributing UFO sightings to "swamp gas".

While it is true that the Dexter event occurred in and around a marshy area, it is certainly curious, even reading news reports of the period, how Hynek, an academic by profession, could approach the science behind his own swamp gas explanation so loosely so as to apply it to any of the reported UFO events of the time, including the events reported at Dexter, Michigan.

The swamp gas hypothesis, according to Hynek's own explanation, stems from plant material that can be trapped under layers of ice and snow during the winter months. As this material decays, gas builds up, and, in some cases, combusts.

To Hynek's credit, he is quoted by the Associated Press in the March 26, 1966 *Dayton (Ohio) Journal Herald* as saying that his swamp gas theory applied only to those sightings (Ann Arbor, Hillsdale, Dexter, Michigan) and not to other reports originating elsewhere in Michigan or around the country.

Compare Dr. Hynek's official explanation at the time to reports made by firsthand witnesses to the events, and one quickly comes to the conclusion that – empirical scientist though he may have been, Dr. J. Allen Hynek was describing something completely different than what was being reported by witnesses to the initial event.

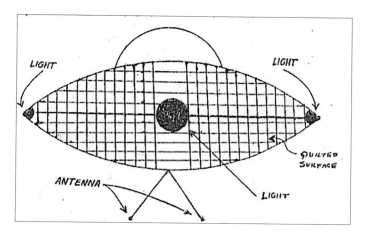

Sketch of the Dexter, Michigan object, as published in the Detroit News
(Courtesy Gerald R. Ford Presidential Library)

Dr. Hynek and the Michigan sighting cases

One of the many reasons why the 1966 Michigan UFO wave is memorable is the fact that at least one of the cases, the March 20, 1966 sighting at Dexter, Michigan, was personally investigated by Dr. J. Allen Hynek, then a scientific consultant to the US Air Force's Project Blue Book.

During his tenure with Project Blue Book, Hynek was, for want of a better term, the project's public face, at least where the media was involved. While Blue Book was headed by Air Force Major Hector Quintanilla, it was often Dr. Hynek who made public statements on behalf of the project to members of the media.

Dr. J. Allen Hynek, discussing the Milan, Michigan case with the news media. Hynek famously dismissed the Milan sighting as the planet Venus.
(Courtesy BETTMAN/CORBIS)

In a very real sense, as stated previously, J. Allen Hynek *was* Project Blue Book. Despite the accomplished military personnel associated with the project such as Major Quintanilla, the media was directed, almost exclusively, to Dr. Hynek, who had served as scientific consultant to the Air Force and, by extension, Projects Sign, Grudge, and Blue Book since 1948.

The Michigan cases, in the opinion of this writer, seem to epitomize the well-reasoned arguments (admittedly in the minority at the time) for the continued study of the UFO subject by both the Air Force, and by extension, mainstream science.

Hynek's own views shifted dramatically over time, earning him the title of "the grandfather of Ufology". Hynek, ever the empirical scientist, consistently argued that UFOs be treated as any other "problem" in science – insisting that one must follow the evidence above all else.

It is the opinion of several researchers in the field – myself included – that the sightings of anomalous objects over Milan and Dexter, Michigan in the spring of 1966 are of vital importance in the overall chronology of the phenomenon, specifically as forensic evidence cases, due to the fact both cases were reported by highly credible sources, the Milan case contains photo evidence, and, perhaps most importantly, in the opinion of this writer, the "explanation" offered for public consumption by Dr. Hynek, the official scientific voice of Project Blue Book, did not seem to conform to the facts of the case, instead offering the broadest interpretation of the "facts in evidence", so to speak, rather than offering a true scientific inquiry into the events in question.

Congressman Gerald Ford and the Michigan sighting cases

Perhaps the most interesting player in the story of the 1966 Michigan UFO wave is not a witness, researcher, or member of the news media, but a politician – the unassuming congressman from Michigan's fifth district, future President Gerald R. Ford.

At the time of the sightings in his home district, Ford held considerable power in the House of Representatives, serving as the House Minority Leader in the 89th Congress (1965-'67). Known for his ability to work across the aisle on various issues, Ford took a keen interest in the numerous UFO sightings reported in his home district, which included the cities of Grand Rapids and Ann Arbor, Michigan. The reason for Ford's interest in the topic could be simple curiosity, or it could stem, at least in part, as is this writer's opinion, from the widely-reported accounts of unknown aircraft sighted over Washington National Airport and the District of Columbia in 1952, during Ford's second term in the House.

As House Minority Leader in 1966, Gerald Ford pushed for Congressional hearings
on the UFO subject.
(Courtesy Gerald R. Ford Presidential Library)

[No. 55]

UNIDENTIFIED FLYING OBJECTS

————————

HEARING

BY

COMMITTEE ON ARMED SERVICES

OF THE

HOUSE OF REPRESENTATIVES

EIGHTY-NINTH CONGRESS

SECOND SESSION

————

APRIL 5, 1966

[Pages of all documents printed in behalf of the activities of the House
Committee on Armed Services are numbered cumulatively to
permit a comprehensive index at the end of the Congress. Page numbers lower than those in
this document refer to other
subjects.]

U.S. GOVERNMENT PRINTING OFFICE

50–066 O   WASHINGTON : 1966

As Minority Leader, Ford was instrumental in convening hearings on UFOs before
the House Armed Services Committee.

(Courtesy Gerald R. Ford Presidential Library)

Rep. L. Mendel Rivers (D-SC), Chair of the House Armed Services Committee, presided over the Congressional hearing into UFOs.

(Courtesy Library of Congress)

Rep. George P. Miller, (R-CA), Chair of the House Committee on Science and Aeronautics.

(Courtesy Collection of US House of Representatives)

The 1966 Michigan sightings in historical context

Perhaps the most important and most notable feature of the 1966 Michigan UFO events is the fact that these cases captured not only the attention of local media, but, in fact, rose to national prominence when they were brought to the attention of Michigan congressman Gerald Ford.

While it was Congressman Gerald Ford who brought what Allen Hynek termed *"the UFO problem"* to the national consciousness with unprecedented hearings before the House Armed Services Committee, it would be President Gerald Ford who would, however indirectly and presumably unintentionally, see that records of the 1966 Michigan UFO wave would be preserved indefinitely.

The documentation of the 1966 Michigan UFO incidents is preserved for posterity due to Gerald Ford's ascension to the presidency upon the resignation of Richard Nixon in 1974.

At the time of Nixon's resignation as a result of the Watergate scandal, any papers created during the president's term of office were considered personal property and not property of the United States government, by way of the Office of the President of the United States, as is the case today.

Fearing that a disgraced president (Nixon or perhaps some future administration) would attempt to conceal misconduct in office by claiming any written material as personal property, Congress passed the Presidential Recordings and Materials Preservation Act of 1974, which was signed into law by President Ford.

While the intention of this Act was to preserve materials related to the abuse of governmental power, it was the forerunner of the Presidential Records Act of 1978, which changed the ownership of presidential materials from personal (personal property) to public (property of the United States government). This is somewhat ironic as the first presidential administration governed by the 1978 act was the administration of President Gerald Ford, whose presidential library was dedicated on April 27, 1981.

It is curious that documentation concerning the 1966 cases can be found at the Gerald R. Ford Presidential Library, as the material dated from his time in the House of Representatives, a full eight years prior to Ford's confirmation as Vice President of the United States under the auspices of the Twenty-Fifth Amendment.

Even more surprising than Ford's preservation of the 1966 material is the sheer volume of said material. Clearly delineated in the Ford collections from the middle 1960s are various press releases, letters to congressional colleagues, and a surprisingly large archive of reference material.

When the Ford Presidential Library provided the aforementioned material to this writer, it totaled some 161 pages.

One of the most ironic documents found amongst the Ford reference material is a scholarly paper authored by well-known University of Arizona physicist and UFO researcher Dr. James McDonald.

In the header, Dr. McDonald writes:

"Mr. Ford: Your 1966 efforts to get the UFO problem clarified appear to have led to only very slight progress. USAF continues to yawn. — J. E. McDonald"

Handwritten note to Congressman Gerald Ford from noted UFO researcher Dr. James McDonald
Courtesy Gerald R. Ford Presidential Library

Thanks in part to the unique confluence of circumstances that saw Ford ascend to the presidency, a rather substantial and copiously detailed UFO case investigation is preserved for posterity under the conservatorship of the National Archives and Records Administration.

Given that, by the National Archives own estimates, *"only 1%-3% [of government documents] are so important for legal or historical reasons that they are kept by us forever"*, UFO researchers should consider themselves exceptionally fortunate that Gerald Ford took such an active interest in the UFO events of 1966.

Without Ford's deeply personal interest in the matter while representing Michigan's fifth congressional district, the UFO subject would certainly have never found its way into a committee hearing before the House of Representatives, let alone archived for posterity as part of a presidential library collection.

It is curious that Ford would, in the years since his presidency, contradict his own previous position with regard to a UFO topic.

The fact remains that Ford did, in fact, spearhead congressional hearings into the UFO topic in 1966 and that the transcript of said hearings is in the possession of the Gerald R. Ford Presidential Library in Ann Arbor, Michigan.

10. Falcon Lake, Manitoba, Canada

The incident at Falcon Lake is perhaps one of Canada's most researched UFO cases. While not a photographic evidence case directly, it does deserve consideration alongside some of the more well-known photographic cases in the field of ufology.

While the photographs available pertaining to the Falcon Lake event have less to do with the event itself as opposed to its aftermath, the Falcon Lake, Manitoba case is valuable from a photographic standpoint, in that the photographs are bona fide examples of evidentiary photography in a police investigation.

On the morning of Friday, May 19, 1967, Stefan Michalak, a Polish immigrant to Canada and an amateur prospector and geologist, went to Falcon Lake from his home in Winnipeg to pursue his passion. He took a bus to Falcon Beach and stayed in the local motel.

The next morning, Michalak packed a lunch and his tools and set off for his day. After several hours of prospecting and studying the local rocks, he sat down to eat his lunch and have some coffee.

Some time during the noon hour, he noticed two cigar-shaped crafts approaching from a south south-west declination. It is important to note that Michalak never referred to the crafts as flying saucers or UFOs, but considered them advanced conventional aircraft.

As he watched the craft approaching, Michalak realized that it was more oval-shaped, inconsistent with military or commercial aircraft of the period.

Still watching the craft's approach, Michalak noted that one object hovered, the other landing on the square top of a rock formation, which he estimated to be approximately 150 feet away from him.

The landed object was first described as being bright red or scarlet, but it soon changed to a light gray, then steel gray and silver, but finally became a light brassy yellow color with a smooth surface.

A few minutes after the object landed, Mr. Michalak noticed a "hatch" had opened, emitting a pale purple light that hurt his eyes. This light was so bright that Michalak needed to put on his prospecting goggles with green filtered lenses. He had not taken a camera on this particular trip, but he did have paper and pencil, so he made a sketch of what he saw.

At this point, it occurred to Mr. Michalak that the object was either part of the US space program or an experimental aircraft. It is again important to note that Michalak did not consider the craft to be a flying saucer or UFO in the commonly accepted sense of the term. He proceeded to approach the object.

By his own account of the incident, Mr. Michalak said he heard child-like voices coming from the object.

Fearing the occupants may be in distress, Michalak, himself a mechanic, called out in English *"Hey, Yankee boys, having trouble? Come on out and we'll see what we can do about it."*

There was no response. Not wanting to leave potentially injured people in harm's way, Michalak called out again several more times in Russian, Ukrainian, German, Italian, French, in his native Polish. There was still no response.

Michalak's curiosity led him to look inside the craft, still wearing his green-filtered goggles. He proceeded to peer inside the craft, later describing the inside of the craft as a maze of light beams going both horizontal and diagonal directions, also noting a series of flashing lights. He believed the beams and flashing lights were operating on a random fashion. He estimated the walls of the craft to be approximately eighteen to twenty inches thick. Mr. Michalak then noticed that two panels closed over the hatch opening and a third panel covered the first two. He also saw what appeared to be a grill pattern on the side which he assumed to be a kind of ventilation system, with each opening approximately three-sixteenths of an inch in diameter.

Still curious, he touched the side of the craft which felt like stainless steel and which was hot to the touch. He noticed his glove had burned. This was one of the gloves he used to chip at the stones. The craft then tilted and took off. As it did so, a blast of hot air from the grill system hit Mr. Michalak and set his shirt and undershirt on fire. He tore the clothing off and saw the craft taking off at an incredible speed.

Falcon Lake as a forensic evidence photo case

Forensic science is defined as the application of a broad spectrum of sciences to answer questions of interest to the legal system. But, besides the relevance to the underlying legal system, forensics encompasses the accepted scholarly or scientific methodology and norms under which the facts regarding an event, or an artifact, or some other physical item are to the broader notion of authentication whereby an interest outside of a legal forum exists in determining whether an object is in fact what it purports to be, or is alleged as being.

The Falcon Lake incident was certainly not a case for the legal system; however, the trace evidence may be of interest to the broader field of ufology for comparison to evidence found in other UFO cases.

The Falcon Lake UFO event presents a wealth of information that may be of interest to researchers in various areas of the phenomenon. In this one event, we, as researchers, are presented with an alleged landing site some fifteen feet in diameter with burned vegetation inside. We have the alleged site testing positive for radiation, to the point where scientists were concerned enough to consider the closing the site to the public. We have multiple burned clothing items and steel tape, all testing positive for radiation. Lastly, of course, we have the grid-like burn on Mr. Michalak's chest and stomach areas.

So, in Falcon Lake, we have a case that is indeed valuable from a forensic point of view. The items mentioned above were subjected to gamma ray spectral analysis and were found to be radioactive. The same was found as a result of tests done on the burned shirt, undershirt, cap and steel tape. These items were initially in the possession of the Defence Research Board (DRB) but were provided to the RCMP Crime Laboratory. It is unclear if the DRB subjected the items to any testing as the DND investigatory results were not found (except the memo which concluded they could not prove the case as false).

The photographic evidence of the grid-like burns on Michalak's chest and stomach are also valuable. They prove that something happened to Michalak that day. Something burned his gloves, hat, shirt and undershirt and set them on fire, resulting in his injury and subsequent illnesses; the burns would reappear on his body approximately every three months.
Does this evidence prove he saw and was injured by a UFO? Indeed it does not. However, the evidence is quite strong that something anomalous did happen.

The purported UFO landing site, Falcon Lake, Manitoba, 1968.
Courtesy Chris Rutkowski / Ufology Research

Applying medico-legal methodology to the Falcon Lake event

Within minutes of the craft's departure, Mr. Michalak fell ill with nausea, several minutes of vomiting, and burns on his chest and stomach. He walked out of the site to the Trans-Canada Highway and managed to flag down a member of the Royal Canadian Mounted Police (RCMP) and related his story to the officer. Michalak returned to the motel and sought assistance to find a doctor, but being a remote area, no medical help was available.

Mr. Michalak returned to Winnipeg where his son took him to the Misericordia Hospital, where he was held for observation. He told the physician he was burned by aircraft exhaust. Mr. Michalak's own physician examined him a couple of days later and concluded the first-degree burns on his stomach and chest area were not serious. No radiation was found. However, Mr. Michalak remained ill for a number of days, and had recurring medical problems for a period of years thereafter.

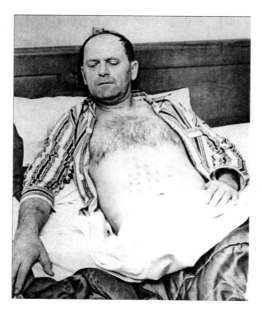

Stefan Michalak, showing circular burns on his chest after his encounter with an unknown object at Falcon Lake, Manitoba.
Courtesy Chris Rutkowski / Ufology Research

The Royal Canadian Mounted Police investigation

An interesting point, with regard to the Falcon Lake Incident, in contrast to cases occurring in the United States, is the fact that an official investigation was conducted by no less than the Royal Canadian Mounted Police. The Falcon Lake Incident, in fact, bears out a rather detailed official investigation on the part of the authorities.

Perhaps the most interesting point, at least in the opinion of this writer, is the fact that, independent of one another, both the Royal Canadian Mounted Police and the Canadian Department of National Defence were unable to disprove Stefan Michalak's version of events, the DND saying essentially this in its official summary, which is, as of this writing, publicly available from Library and Archives Canada.

While this may not be the proverbial smoking gun in the history of ufology, it most certainly highlights key differences between the handling of UFO cases on the governmental level between the United States and Canada. This case also offers researchers a unique opportunity, as Library and Archives Canada makes what UFO material it has publicly available via the Internet.

UFO REPORT
FALCON LAKE, MAN.

A Mr. Steven Michalak of Winnipeg, Manitoba reported that
he had come into physical contact with a UFO during a prospecting
trip in the Falcon Lake area, some 90 miles east of Winnipeg on the
20 May 67. Mr. Michalak stated that he was examining a rock formation when two UFOs appeared before him. One of the UFOs remained
airborne in the immediate area for a few moments, then flew off at
great speed. The second UFO landed a few hundred feet away from
his position. As he approached the UFO, a side door opened and
voices were heard coming from within. Mr. Michalak states he approached the object but was unable to see inside due to a bright
yellow bluish light which blocked his vision. He endeavoured to com-
municate with the personnel inside the object but without result. As
he approached within a few feet of the object, the door closed, he
heard a whining noise and the object commenced to rotate anti-clockwise
and finally raised off the ground. He reached out with his left
gloved hand and touched the object prior to its lifting off the ground;
the glove burned immediately as he touched the object. As the object
left the ground the exhaust gases burned his cap, outer and inner
garments and he sustained rather severe stomach and chest burns. As
a result of these he was hospitalized for a number of days. The doc-
tors who attended and interviewed Mr. Michalak were unable to obtain
any information which could account for the burns to his body. The
personal items of clothing which were alleged to have been burnt by
the UFO were subjected to an extensive analysis at the RCMP Crime
laboratory. The analysis was unable to reach any conclusion as to
what may have caused the burn damage. Soil samples taken from the immediate area occupied by the UFO by Mr. Michalak were analysed and
found to be radioactive to a degree that the samples had to be safely
disposed of. An examination of the alleged UFO landing area was
tested by a radiologist from the Department of Health and Welfare and
a small area was found to be radioactive. The Radiologist was unable
to provide an explanation as to what caused this area to become con-
taminated.

Both DND and RCMP investigation teams were unable to provide
evidence which would dispute Mr. Michalak's story.

DND summary of the Falcon Lake Incident. It is important to note the closing
statement of the report.
Courtesy Library and Archives Canada

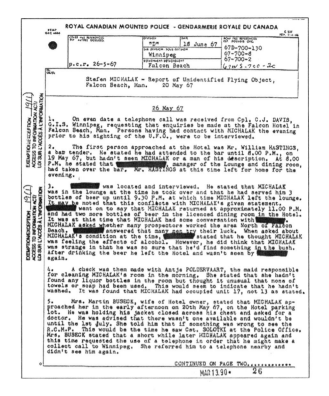

RCMP Report on the Falcon Lake Incident
Courtesy Library and Archives Canada

The Department of National Defence

Both the RCMP and DND initial involvement in this case began when Mr. Michalak left the area he was exploring and flagged down a patrol car on the Trans-Canada Highway. The Constable's report reiterated Mr. Michalak's story but the Constable thought that Michalak had been drinking and was now in a hangover situation, as Mr. Michalak had red eyes and was acting irrationally. Michalak then showed the Constable the burn marks on his chest and stomach, however, the RCMP member believed he had rubbed ashes on the area.

A few days after the incident, two RCMP members visited Mr. Michalak at his home in Winnipeg. When they arrived, there was a member of the Aerial Phenomenon Research Organization (APRO) present, but the APRO member left prior to the RCMP interview. Mr. Michalak stood firm to his story. Subsequent interviews also included members of the Royal Canadian Air Force (RCAF).

A search for the landing site was initiated by the RCMP and DND but they were unable to find the area. A few days after the initial search, the RCMP and DND flew over the area, but Michalak was unable to locate the sport from the air. However, he later travelled to Falcon Lake with a friend and he was able to locate the site, which they photographed.

The photograph showed the alleged landing site to be about 15 feet in diameter with the vegetation burned in the area. Subsequently, the Department of National Health and Welfare (now Health Canada) sent a radiologist to the site for testing. The radiologist found the site to be radioactive to a degree which caused the Department to consider restricting use of the area. Other materials found at the sited, including a steel tape measure were found to be so radioactive that safe disposal was required.

National Defence Headquarters, Ottawa, Canada
Courtesy Mike Powell / Wikimedia Foundation

Investigations by both the RCMP and DND were unable to conclude that Mr. Michalak's story was false.

The Canadian Government's official position

As in other countries, the Canadian Government no longer has any official interest in the investigations of reported UFO sightings.

Prior to 1967, UFO investigations were the responsibility of the Department of National Defence. In a letter dated September 1967, DND wrote to the Chairman of the Advisory Committee on Scientific and Industrial Research requesting that the responsibility of investigating UFOs be transferred to the National Research Council (NRC) where scientific research facilities and trained personnel were available for carrying out an objective investigation.

When a UFO report was referred to DND for investigation, they were studied by the Operations staff who would endeavor to classify the information into one of two categories:

Category one – Information which would suggest this type of phenomena associated with fireballs and meteorites, or

Category two – Information which did not conform to the physical patterns usually associated with fireballs or meteorites. Category one reports were forwarded to the NRC Meteorite Centre for scientific study.

Category two reports were placed on file and annotated that nor further action was required, or action was initiated to conduct a formal investigation of the report by a military officer.

What is important in determining the official position of the Canadian Government with regard to UFOs is stated directly in this letter from DND. It states,

"Investigations conducted to date have failed to disclose any evidence which would suggest that UFOs pose a threat to national security. However, a number of investigations suggest the possibility of UFOs exhibiting some unique scientific information or advanced technology which could possibly contribute to scientific or technical research. From information available on UFO activity it would appear that the primary interest lies in the field of scientific and technical research and to a lesser degree to one that poses a threat to national security."

To this writer, it appears that the Canadian Government has admitted to the existence of Unidentified Flying Objects.

Of all the files subsequently transferred to the NRC, three were considered to be 'unsolved'. One of the files was the Falcon Lake incident. The other two files in question, while not part of the study in this book, but have been covered at some length elsewhere. They are the Duhamel, Alberta crop circles and the incident at Shag Harbour, Nova Scotia. As a matter of clarification, prominent Canadian ufologists Chris Styles and Don Ledger have authored a book, called *Dark Object*, on the Shag Harbour Incident.

In the 1980s, the NRC discontinued any further investigations into UFOs and the files were transferred to the National Archives (now Library and Archives Canada). As is the case in most countries, the National Archives is not an investigative body, but rather a repository for Canada's important historical records. UFO reports received at Library and Archives Canada are now forwarded to Chris Rutkowski of Ufology Research of Manitoba (UFORUM), who compiles an annual listing which is posted on the UFORUM website. One interesting item of note: Rutkowski has seen an approximate 10% increase in sightings in Canada, the reasons for which are unknown.

The bottom line, in my view, is this: Canada appears to acknowledge the existence of UFOs. Canada has not, however, claimed that UFOs are interplanetary or extraterrestrial in nature. While the available forensic evidence does not offer proof as to the origins of the UFO phenomenon, the evidence does seem to suggest that something strange is indeed happening.

11. The Yorba Linda Photograpy

The ever advancing field of photography – particularly in the digital age – is, at its very core, about conveying a message through a digital medium.

While this is best accomplished using the clearest, brightest, highest resolution images available, there are instances where the best image available is, in fact, the only image available.

The January 24, 1967 sighting of a UFO in the skies over Yorba Linda, California is just such an instance.

According to all accounts, the evening in question was overcast, dreary, and generally uneventful – until a chance encounter by a teenage boy gave the field of ufology what is perhaps one of the more interesting cases of photographic evidence, though not necessarily for the reasons most commonly thought.

Some photographic UFO cases are intriguing due to their detail and clarity. The Yorba Linda photograph is intriguing, perhaps, given the absence of these details, among other factors.

The Yorba Linda photograph, while technically a flawed image, offers researchers enough detail to formulate reasonable hypotheses, and therefore, deserves consideration as yet another piece of anecdotal evidence toward the argument that at least a percentage of UFO reports are intelligently controlled craft which operate under an alternative understanding of physics.

Yorba Linda as representative of the majority of UFO photo cases

In the opinion of this writer, the Yorba Linda photograph is very much a representative example of the majority of photographic cases on record in that, with few notable exceptions, the issues of image clarity and resolution are common impediments to investigation.

Simply put, very few photographic cases on record today yield clearly defined images of a purported UFO. The overwhelming majority of images, while perhaps not substandard in quality, certainly require some level of digital restoration, for want of a better term, before an investigation can begin in earnest.

As the term "restoration" is, in itself, I feel it is important to properly define it, particularly when the restoration being undertaken is for an investigative purpose.

Given the power and versatility of photo editing software packages such as Adobe Photoshop, the tools required to alter witness-submitted images are certainly readily available, however, it is important to adhere to generally accepted standards, where digital images are concerned.

In the interests of brevity and clarity, generally acceptable uses of image enhancement software in the fields of law enforcement and journalism can also be applied to the field of ufology, as the intent is simply the enhancement – not the alteration – of data.

Image enhancements commonly accepted both in the field of journalism and courts of law include, for example, increasing image brightness, contrast, or other inherent properties for purposes of clarity, while maintaining its study was evidence.

These image enhancements are necessary – in a majority of cases – due to the fact that, as mentioned previously, few, if any, anomalous photographs are captured under what could be considered ideal conditions.   The Yorba Linda photograph is representative of countless additional questioned photos captured to date, specifically because it was captured under relatively ordinary conditions, allowing for additional testing.

In very simple terms, the Yorba Linda photograph is a remarkable piece of photographic evidence, in the opinion of this writer, due to its relatively unremarkable origins.

Imperial Mark XII camera
Courtesy CameraPedia.org
via GNU Free Documentation License

The Unlikely Witness

In the criminal justice arena, absent an identifiable suspect, suspicion almost immediately falls on those individuals closest to the center of the investigation, specifically witnesses who have come forward or, in the case of assaultive offenses, those most intimately acquainted with the victim.

As the Yorba Linda photo case leaves investigators with only two viable "persons of interest", it is certainly reasonable to examine whether or not the witness/photographer could be responsible for perpetrating a hoax, rather than documenting an actual event.

The Yorba Linda photo case is unique and is noteworthy perhaps for its simplicity. This case deserves the attention of researchers for the very simple reason that the hoax hypothesis is seemingly discounted from the very beginning. This is due to the fact that the initial witness in this case was a teenager at the time of the event and, while this is not to say that a teenager would not have the motivation to perpetrate a hoax, it is reasonable to assume that a fourteen-year-old, despite an apparent interest in photography, would lack the skill required to hoax a UFO photograph, even one as relatively nondescript as the Yorba Linda photograph.

Taking into account the limited background information available on the witness, it is difficult to rule anything out definitively, however, considering what is known, it is hard to imagine that there would be any motivation whatsoever for a highly intelligent, apparently honest, well-liked fourteen-year-old to fabricate a photograph of a UFO, especially when the witness had previously considered the subject to be "silly and uninteresting".

What is particularly interesting is the fact that the January 24, 1967 incident was the witness's second such experience in less than three weeks.

This is certainly curious, but based on the information available, it seems purely coincidental that the witness would first see an anomalous object on January 4, 1967 and photograph a similarly anomalous object less than three weeks later.

It is also important to consider the scrutiny that anyone reporting a UFO experience must inherently subject themselves to in going public. This case was investigated by veteran ufologist Ann Druffel, then of NICAP, who would remain involved with this case for over four years. In that time, Druffel would refer the teenage witness to no less than half a dozen other researchers, all of whom were professional scientists.

Even adult witnesses have wilted under this sort of pressure, but this witness faced these critics head on, each individual investigator judging this young man to be credible.

As an objective observer, coming to the case decades after the fact, I was initially struck by the phenomenally young age of the witness at the time of the event, as well as his commitment to the investigation over a period of years, as he was referred to photographic expert after photographic expert. While the sighting and subsequent photograph are intriguing in their own rights, it is equally astounding, in my view, that a witness barely of high school age could find within himself the fortitude to commit himself to years of cross-examination by professional scientists who had dedicated their free time to resolving what Allen Hynek termed "the UFO problem".

Technical factors support the 'genuine event' hypothesis

The Yorba Linda photo case is, above all else, a truly unique example within the field of ufology, as the photograph itself is a textbook example of forensic evidence, the likes of which are typically confined to the field of law enforcement.

While the events in other UFO cases can be ascertained to within a high degree of probability, the singular photo taken during the Yorba Linda event can be linked to the scene and the photographer forensically – much the same way that a criminal defendant can be linked to a crime scene – to the exclusion of all other possible scenarios.

While the photographer in this case, now grown, has never altered his original statement, to this writer's knowledge, the fact that the Yorba Linda photograph can be forensically linked to a camera in the witness's possession strikes a serious blow to the stock argument of debunkers of this phenomenon – that UFO researchers practice 'bad science'.

In the case of the Yorba Linda object, researchers can definitively say that the questioned photograph was produced by the witness's Imperial Mark XII camera – to the exclusion of all other Mark XII cameras – based on a damaged winding mechanism, which produced the scratches visible on the photographic negative and, consequently, the resulting print.

While evidence that would allow researchers to identify the Yorba Linda object – or to state categorically its origin – does not exist, science has offered the research community enough information to show that the object in the photograph, as yet unidentified to the best of my knowledge, was, in fact, a physical object occupying that space in the night sky at that particular moment in time and was not the product of a hoax.

Another piece of evidence which bolsters the credibility of both the witness and the case overall, in my opinion, is the fact that this incident represents a photographic case with an unbroken chain of custody – the film, negative, and resulting photograph remained in the possession of the witness at all times due to the fact that the negative was not professionally developed.

The fact that the negative and print were both processed in a home darkroom, rather than in a professional setting, may also account for the overly dark, somewhat uneven appearance of the print. These effects could result from a number of factors, including an improperly set enlarger, over immersion in developing chemicals, expired developing chemicals, or an insufficient amount of time in a fixing bath, which would cause the print to darken excessively once exposed to natural light.

While any one cause could potentially create a photographic print exhibiting the properties of the Yorba Linda photograph, it would seem more likely, in my opinion, that the condition of the print was the result of a combination of factors, rather than a single cause.

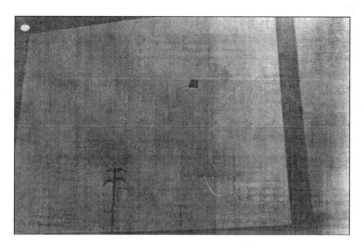

The Yorba Linda UFO Photograph in its original state
Courtesy Fund for UFO Research, Inc.

The Fund for UFO Research Analysis

While not a research organization per se, the Fund for UFO Research (FUFOR) serves a vital function within the field of ufology, providing research grants to individual researchers or other organizations on a case-by-case basis to carry out their research.

With regard to the Yorba Linda photographic case, the Fund for UFO Research was instrumental in bringing the case back into the public eye with the publication of *UFOs Exposed: The Classic Photographs* by veteran ufologist Robert Swiatek in 2004.

Though not an exhaustive work, *UFOs Exposed* serves a vital function, in offering overviews of a great many cases, Yorba Linda included, while supplying all the facts pertinent to the case, allowing the reader to form a reasonable conclusion – the very definition of a circumstantial evidence case.

FUFOR's contribution, by way of author Robert Swiatek, cannot be overstated, in that the ever-brief overview of the Yorba Linda case goes back to the original case material, drawing from original investigator Ann Druffel, as well as technical data, which includes simplifying photographic analysis and details of the printing process, allowing them to be condensed and easily understood by the general reader.

It is highly important, in the opinion of this writer, to note the qualifications of the author of *UFOs Exposed* when referencing the Yorba Linda case because, although not directly involved with the investigation at the time, veteran ufologist Robert Swiatek is highly qualified to address some of the features that present themselves in the Yorba Linda photograph, as well as similar photographs.

Though not a photographic expert, Swiatek is, in this writer's estimation, certainly capable of rendering an "expert opinion" on the case, given his background, which include degrees is both physics and earth sciences. Both of these disciplines come into play when studying the admittedly poor quality Yorba Linda photograph, in that the object's size, distance from the camera, and altitude can all be approximated by studying the print and making observations regarding the unknown object relative to other objects present in the image.

Digitally rendered version of the Yorba Linda Photograph, which was used as the cover image for *UFOs Exposed.*

Courtesy Fund for UFO Research, Inc.

The Yorba Linda photograph in historical context

The field of ufology, like any field of science, is constantly evolving. The scientific inquiry into what Hynek termed "the UFO problem" is continuously being redefined by precedent-setting discoveries made through fieldwork and the reevaluation of cold cases.

When evaluating photographic cases in order to determine their place within the history of the phenomenon, the question that springs to mind is this:

"Had this case never come to light, how might our understanding of the UFO problem be different?"

While the answer to this question might not be immediately apparent, with respect to the Yorba Linda event, the case is nonetheless important when several factors are taken into consideration.

Chiefly, the Yorba Linda UFO event is significant in that it was, perhaps, the first if not the only single-witness case on record involving a minor.

The age of the witness might seem immaterial at first blush; however, it speaks directly to the witness's credibility. While detractors will argue that a fourteen-year-old is too young to be considered a reliable witness, the fact remains that this witness continued to meet with photographic experts for four years following the initial event, passing muster each time.

The case is also of particular interest, as mentioned previously, due to the fact that it offers researchers an uninterrupted chain of custody. This is vitally important when one applies the standards of forensic science to UFO field investigation. The chain of custody issue has become increasingly important as case investigations have become more complex with the passage of time, often requiring the expert opinions of multiple investigators.

In comparison to other cases of the period, the Yorba Linda photo case may seem unlikable, but the fact remains that the case is as compelling today as it was over forty years ago. The fact that this case has been reinvestigated over half a dozen times since the initial report – with standing scrutiny each time – should speak volumes toward its inclusion alongside some of the best-known UFO photos to date.

12. The Colfax Wisconsin Photograph

The village of Colfax, Wisconsin is an exceedingly small town, even by the most generous of definitions. Boasting a total area under one and one-half miles and the population of barely 1,100 people, Colfax is located in Dunn County, some twenty-three miles from the city of Eau Claire. It seems one of the unlikeliest locations for a UFO sighting, yet this is exactly what happened on the afternoon of Monday, April 19, 1976.

That day, local police officer Mark Coltrane was on routine patrol when he stopped for lunch break. While tending to a malfunctioning radio, Officer Coltrane witnessed what can only be described as an unconventional flying object. Officer Coltrane proceeded to snap two photographs of the unidentified craft using his department-issued camera.

The fact that Coltrane's police radio was malfunctioning is significant, as a great many sighting reports throughout the years have reported similar instances of electronic disruption.

Colfax, Wisconsin as a high credibility case

The Colfax, Wisconsin case, in the opinion of this writer, must be considered a high credibility case for the very simple reason that there seems to be no valid argument why the witness in this case, a police officer, would perpetrate a hoax or have reason to fabricate not only a story, but also an accompanying photograph.

On the basis of the image alone, the Colfax case should be considered highly credible, in my opinion, because, regardless of what was photographed, the images in question were and are unable to be altered.

Though not an ideal medium, the Polaroid print does offer a few advantages over conventional photography, due to their unique developing method.

Taking this unique developing chemistry into account, Polaroid prints are virtually impossible to hoax or older by conventional means.

Additionally and perhaps more importantly, this developing process provides, in a very real sense, chain of custody for that one particular image from the moment the photograph was taken.

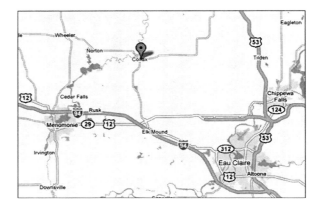

Map showing the location of Colfax, Wisconsin
Courtesy of Google Maps

Police and other professionals as witnesses to anomalous events

Police officers, as well as others in the law enforcement profession, are perhaps among the best potential witnesses to normal civilians such as a UFO sighting.

Law enforcement training is specialized, in that these professionals are trained to be expert witnesses. The very nature of these professions requires the ability to recall details that might escape a civilian witness.

This is not to say that civilian witnesses are any less important or their observations any less valid. As with any degree of training, certain abilities and skill sets are being honed and developed for a very specialized purpose. It is for this reason, in the opinion of this writer, that certain professionals – law enforcement officers especially – are highly valuable witnesses to events such as the Colfax UFO case.

A cursory analysis of the Colfax photos

Very little firsthand data is available regarding the Colfax event, although certain general observations can be made, providing a very cursory analysis of the questioned photograph.

Given that the original photograph was taken by a police officer using a police-issued Polaroid camera, the issue of chain of custody and authenticity is immediately established. In the opinion of this writer, the Colfax case shares a number of similarities with the far better known Heflin incident.

Photographic analysis is a bit more difficult when dealing with Polaroid images, such as the Heflin and Colfax photographs, given the fixed image size and lack of a photographic negative. Prints from Polaroid photos must be made directly from a positive image, which, ultimately, leads to a certain degree of degradation in the second-generation copy.

In addition to distortion or loss of quality through the duplication process, any enlargement to a second-generation (or later) copy would cause significant pixilation, thereby limiting, if not precluding, any image enhancement or analysis from being conducted.

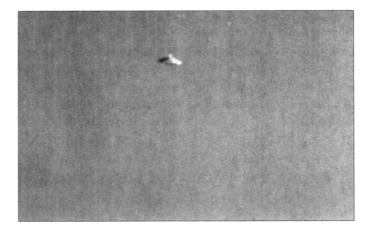

Colfax, Wisconsin Polaroid photo, 1976
Courtesy of BJ Booth / UFOCasebook

The argument against a hoax determination

The argument against labeling the Colfax case a hoax rests in part on the witness's standing in the local community as a member of the police force, but also squarely on the validity of the photographs captured during his initial encounter.

Like the Heflin incident before it, the Colfax case rests exclusively on the validity of Polaroid photographs.

While not particularly suited for photographing objects at any great distance, Polaroid cameras, as evidenced by the Heflin incident, are quite capable of capturing images of fair to moderate quality. Both Officer Mark Coltrane and Rex Heflin relied on Polaroid in their respective careers and seized upon it being readily available when needed.

It is important to note that both Coltrane and Heflin were carrying camera equipment as part of their professional lives, and therefore the equipment would have been on hand at all times. The argument for a hoax in either case, in the opinion of this writer, just does not fit the facts. Both men would stand to lose far more than they would gain by way of publicity.

As referenced previously with regard to the Heflin case, the physical design of the Polaroid camera would virtually preclude the photographer from being able to perpetrate a hoax by way of conventional means, such as double exposure, given that the development of the photographic print happens in real time without the use of a negative.

The Colfax , Wisconsin event as a precursor to other law enforcement reports of anomalous craft

The Colfax, Wisconsin event can be seen as the precursor to other reports of anomalous craft by law enforcement personnel for a variety of reasons.

Firstly, with the notable exceptions of the Milan and Dexter, Michigan sightings of the mid-1960s, very few law-enforcement officers (outside of the US military) involved themselves with UFO sightings at the time.

Despite initial trepidation and fears of reprisal, Officer Mark Coltrane did, in fact, come forward with his experiences. This very same behavior can be seen in other prominent cases as well, most notably the multi-jurisdictional reports which centered around Millstadt, Illinois in January of the year 2000.

Though there may not be any specific connection between cases other than the law enforcement profession, there is certainly an interesting component at play given the law enforcement background.

As 'pillars of the community', police and other law-enforcement officials are, by their nature of their work, imbued with the public trust and held to a standard higher than that of a civilian witness.

While perhaps not the first, Officer Mark Coltrane and the Colfax, Wisconsin UFO case are nonetheless noteworthy for setting a standard, at least to some degree, regarding the UFO subject and witnesses within the field of law enforcement.

13. Vancouver Island, BC

In October 1981, Hannah McRoberts, her husband, and their young daughter took a five-day vacation to the east coast of Vancouver Island, British Columbia during the Canadian Thanksgiving weekend. As a matter of clarification, in Canada, Thanksgiving Day falls on the second Monday in the month of October, directly coinciding with the observance of Columbus Day in the United States.

Like most families on vacation, photographs were taken to preserve the memories of the trip. At one point during their trip, they observed a mountain peak with a cloud around it. The family thought the scene resembled a volcano spewing steam and this prompted Mrs. McRoberts to take a picture of it. None of the family members noticed anything else in the area.

Mrs. McRoberts used a 35mm Mamiya camera with a 50 – 55 mm lens, 125 speed and ASA 100 film. Irrespective of the brand, the camera equipment in question was quite common for the period, was readily available in the consumer market, and is, generally speaking, fairly representative of the type of camera most families might pack when going on vacation.

It is important to note these details, while they may seem fairly obvious to the general public, as the technical limitations of any camera certainly do factor in to the photographs that are produced by said camera.

According to both witness statements and published reports, this incident took place on October 8, 1981 at approximately 11:00 AM. Even without direct witness statements, the time of the event can be roughly approximated, given other factors present in the original image, such as the position of shadows, the position of the sun in the sky, or cloud cover.

Mamiya 528AL, the same model used by Hannah McRoberts
Courtesy Ken Lyndrup

When the prints were received after processing, an object was noted to the right of and above the mountain peak. The photograph was subsequently provided to a staff member of the McMillan Planetarium in Vancouver, who, upon seeing the object, contacted the Aerial Phenomenon Research Organization (APRO). Following a series of analytical procedures were done on the picture, it was given to Dr. Richard Haines, who was then the Editor of the *Journal of Scientific Exploration*, for further analysis. Dr. Haines concluded that Mrs. McRoberts had unwittingly captured a photograph of a genuine airborne object that was not the result of emulsion deformity or optical illusion associated with the camera's inner mechanisms.

The 1981 Vancouver photo compared to other well-known UFO photographs

The Vancouver Island photograph compares favorably, in my view, to several other photographs that have surfaced throughout the history of modern ufology.

As stated by Dr. James Harder in support of Dr. Richard Haines's 1986 MUFON symposium paper, the Vancouver Island photograph appears to be of excellent quality and likely depicts a physical object and not some type of optical artifact. Based solely on the evidence of the photograph itself, it is reasonable to draw the conclusion that the object pictured is a UFO in the most basic definition of the term – an as-yet-unidentified constructed object capable of flight.

Vancouver Island compares favorably to a fair number of cases in the modern era of ufology, perhaps chiefly due to the clarity of the image. In my view, this has more to do with the advances made in photographic technology than anything else.

The 1950 McMinnville, Oregon photographs, for example, are perhaps some of the most iconic in the history of ufology, having been reprinted countless times in the United States alone. While, to this writer, the objects in both photographs appear strikingly similar and it certainly can be argued that, compared side-by-side, they could well be either identical or strikingly similar craft, the Vancouver Island photograph and event are considerably less well-known, despite arguably superior image clarity.

The same can be said for the equally well-known Heflin photographs taken in Santa Ana, California in 1965. As is the case with the McMinnville photographs, the issue here seems to be a matter of image quality and clarity. Advancements in photographic technology between the middle 1960s and the 1980s make a profound difference. One can only theorize what detail might have been captured, and what evidence ultimately gained, had either Paul Trent or Rex Heflin captured their famous images using even 1980s-era technology.

As researchers have said, Vancouver Island represented, at the time, perhaps some of the best photographic evidence available. There is no doubt in the mind of this writer that there will be even more fantastic photo cases that will present themselves in the future, however the Vancouver Island photograph, if not the best piece of photo evidence for its time, surely deserves consideration as one of the preeminent cases in recent history.

Vancouver Island UFO Photograph, 1981
Courtesy Fund for UFO Research, Inc.
Image by Hannah McRoberts

Technical analysis of the 1981 Vancouver Island photo

As mentioned previously, the camera was a Mamiya 35 mm, model 528AL, single lens reflex (SLR) type with a permanently attached Mamiya/Sekor 48 mm lens with a 1:2.8 aperture. This particular camera is an automatic, which, as the name implies, means the only things the photographer has to do are load the film properly, set the ASA number for the film, aim, and manually focus the lens. Both the shutter speed and the aperture adjust automatically for the best exposure. The camera has a finger-depressed activation lever and once the shutter has opened and closed, another picture cannot be taken until the film has been advanced to the next frame.

Dr. Haines borrowed the McRoberts' camera and took a series of test photographs under similar sun angle, sky brightness and other similar conditions in order to determine if there were any lens and/or shutter-related image artifacts. He found none. Then lens was coated with standard anti-scattering material. There were no scratches or other flaws found on the lens itself. The only item of note Haines found, with respect to the mechanical workings of the camera was a small scratch on the original negative, far removed from any significant portion of the image. Small scratches such as these are common in film processing, many occurring during the developing and printing processes.

Mrs. McRoberts had used a Kodak Safety Film 5035, 35 mm, most commonly known as Kodacolor II, with an ASA rating of 100. This was what is known as a C-41 process film. Most personally developed print film is developed using the C-41 color development process.

Upon returning from the family's trip, Mrs. McRoberts had the film commercially developed, waiting the normal period of about one and a half weeks to receive her photographs. She did not order any special development processes or enlargements. She received the standard "jumbo" size colored prints (by many processors standards, this would equate to a print size of approximately 4 x 6") and the developed color negatives.

As mentioned previously, it was only after receiving the developed photographs that the strange aerial disc was noticed.

Dr. Haines took the following steps in his analysis of the negative, as well as the first generation positive black and white and color prints:

- Linear and angular measurements

- Microdensitrometry scans (to establish optical density range)

- Black and white photographic enlargements using papers with different spectral sensitivities

- Computer-based contrast enhancements.

In the interest of brevity, Dr. Haines's research will not be discussed in detail here. However, should readers be interested in the technical data behind the analysis of the Vancouver Island photograph, Dr. Haines's research can be found in the proceedings of the 1986 MUFON Symposium, UFOs: Beyond the Mainstream of Science, which was held in East Lansing, Michigan June 27 – 29, 1986.

Having said this, there was something interesting about the photograph, according to Haines. This is that the brightest area in the disc/object/UFO was well under the brightness of the cloud. Dr. Haines noted that according to a recent (as of 1980) optical handbook,

> "Within the visible spectrum, a smooth, polished silver surface reflects increasingly higher percentages of incident radiation with increasing wavelength. Polished aluminum reflects about 85% regardless of wavelength, as does nickel at about 60%, silicon at about 30% and steel at 54%."

Dr. Haines' comparison of darker areas on the negative suggested that the surface of the disc (UFO) was not a polished surface of any of these materials.

Also worthy of note is that Dr. Haines analyzed for a possible double exposure. However, the type of camera used by Mrs. McRoberts prevents this with a fame locking mechanism, that is, the photographer has to physically advance the film before another photograph can be taken, effectively eliminating the possibility of the image being the product of double exposure.

Secondary factors that rule out a prosaic explanation

Debunkers and skeptics have suggested that the object Mrs. McRoberts photographed was that of a "Frisbee" or similar type of toy. Dr. Haines, along with his father and the McRoberts, visited the site at the same time 2 years after the picture was taken. From the site inspection, it was clear that there was enough flat ground for someone to have flown a model airplane or thrown a "Frisbee" into the air. However, Dr. Haines' analysis strongly suggested that the top surface of the object/disc in the photograph intersects the front edge in a relatively sharp contour rather than in a smoothly curved contour. And, "Frisbees" don't have sharply intersecting surfaces.

Dr. Haines found the McRoberts to be honest, hard-working people. He found nothing in their background which would lead him to believe that they had perpetrated a hoax. Both of the McRoberts were puzzled as to the origin of the object/disc/UFO found on the photograph. They were not defensive, nor did it appear they attempted to cover up anything. In their home, there was no evidence of any interest in the occult, the psychic or metaphysical, or other similar subjects. Mrs. McRoberts indicated she had a casual interest in UFOs, but had not read any books on the subject. Both Mr. and Mrs. McRoberts had seen the Hollywood film *Close Encounters of the Third Kind*, which was released in 1977.

Mr. McRoberts had been an avid science fiction reader in his youth and readily admitted to owning a Frisbee, as well as being proficient with it. Dr. Haines inspected the Frisbee and found no evidence it had been tampered with or modified in any way that would allow for it to have been the explanation behind the questioned photograph from Vancouver Island.

Copies of the photograph were given to the Aerial Phenomenon Research Organization (APRO), which, in turn, gave them to Dr. James Harder, then Professor of Engineering at the University of California at Berkeley.

In the United Kingdom, the prints were given to Mr. Percy Hennel, who was generally regarded as Britain's leading expert in all matters related to color photography. Mr. Hennel's conclusion was that he found nothing suspicious about the photograph and he warned that if the picture was genuine, then the disc was enormous – as much as several hundreds of feet wide – in order for it to have shown up so large at such a distance.

Dr. Harder's conclusion was the following: *"All things considered, the photo presented here appears to be an excellent and probably genuine photo of a classical disc photographed in daylight."* Harder further concluded that what Mrs. McRoberts had unwittingly captured in her photo of the mountain was a genuine airborne object and not the result of emulsion deformity or optical illusion, validating Dr. Haines original impressions.

The Fund for UFO Research Analysis

The Fund for UFO Research, as has been mentioned with regard to its work with other cases, is not an investigative body per se, but frequently compiles research and publishes technical reports and scholarly papers detailing research into well-known UFO cases.

With the 2004 publication of *UFOs Exposed: The Classic Photographs*, the Fund for UFO Research named the Vancouver Island photograph as being one of the better photographic cases in the history of the phenomenon.

Though by no means a complete account of the event, the case abstract, as presented by physicist and veteran ufologist Robert Swiatek, certainly provides the community at large with the working knowledge of the case needed to form the most basic of hypotheses.

The Fund for UFO Research has, at its disposal, individuals who are not only highly respected within the field of ufology, but also within their respective careers and scientific disciplines. It is for this reason, in my opinion, that while *UFOs Exposed* offers a very basic overview of each case profiled, each of the case abstracts contained therein can be viewed as expert testimony.

The Vancouver Island, British Columbia incident in historical context

The Vancouver Island, British Columbia incident is, in the opinion of this writer, a vitally important piece of the proverbial puzzle as ufology advances forward.

As previously mentioned, the advances in photographic technology set the Vancouver Island photograph apart from earlier cases, while cases that postdate the Vancouver Island event continue the same trend.

Just as the Heflin photos defined an earlier era in ufology, the Vancouver Island event was, in the opinion of this writer, equally as iconic a case from a photographic standpoint, prior to the 1987 Gulf Breeze, Florida events.

Simply put, each decade of research has a very select number of cases that can be held up as the best examples from that time period. In this instance, I believe that the photograph captured on Vancouver Island, British Columbia in October of 1981 is not only among the best evidence cases for its time, but as suggested by Rob Swiatek and the Fund for UFO Research, Vancouver Island is an example of perhaps one of the best photo evidence cases yet on record.

14. The Gulf Breeze Florida Photographs

The city of Gulf Breeze, Florida is located in Santa Rosa County, on the Fairpoint Peninsula in the western panhandle.[39]

Gulf Breeze, like Roswell, New Mexico before it, has gone out of its way to embrace the UFO subject and the income to the local economy generated by UFO researchers or others interested in the events that put the quiet Florida town on the national map in the 1980s.[40]

The Gulf Breeze, Florida sightings are an anomaly even within the field of UFO research. The events at Gulf Breeze are unique in that the investigation into the events was being conducted while the events were ongoing. While the official case file is designated as a close encounter of the third kind (CE-III) with photographic evidence involved, I will, in the interest of brevity, only concern myself with the photographic elements of this case.

The commonly held sequence of events begins on Veterans Day – November 11, 1987 – at approximately 5 o'clock in the evening.[41]

Local contractor Edward Walters was at work in his home office when an odd light coming from his yard caught his attention. Minutes later, Walters would snap the first of some forty photographs of unexplained aircraft over the City of Gulf Breeze. Investigations into this and other sightings reported in and around Gulf Breeze, Florida would continue for a period of some two and a half years, though the bulk of the photographs commonly referred to as *"The Gulf Breeze Photos"* span only a six month period, from November of 1987 through May of 1988.

One of the many Gulf Breeze UFO photos, circa 1987

The Gulf Breeze event of early November, 1987 was immediately taken seriously not only due to the photographs captured by the initial witness on the evening of November 11, but also by statements made by at least three corroborating witnesses, all of whom reported seeing the same object that evening. (MUFON UFO Journal # 240, April, 1988)

It should also be pointed out that all of the additional witnesses to this initial event were several miles from one another, the secondary sighting occurring approximately ten miles east-northeast of the location where Walters took the first set of photographs in this case.

The Gulf Breeze photos compared to other well-known UFO photos

The Gulf Breeze events, that is to say the Walters photographs, have become the modern equivalent of the McMinnville, Oregon photos – instantly recognizable to both those interested in the field of UFO research and members of the general public.

The most recognizable photographs from this series include the "blue beam" photograph, which shows, as the name implies, a beam of blue light coming from the underside of the craft, as well as numerous photos shot by Walters at the approximate level of the horizon or in the immediate vicinity of other objects, such as trees, perhaps as a frame of reference.

Like the McMinnville (Trent) photos and Heflin photos before it, the Gulf Breeze photos are important for their inclusion of other objects for use as "landmarks", if you will, to aid in analysis at a later time. In all of the aforementioned cases, the data gathered from background objects – such as the garage in the McMinnville photos or the windshield, window frame, and highway of the Heflin Photos – provide researchers with enough additional information about the time when each respective photo was taken so as to help further shape their hypotheses into what the questioned or unknown object might be.

Gulf Breeze as a 'special circumstance' case

The initial photographs were shot with a Polaroid camera using 108 film, which is equivalent to 35mm film with an ISO of 100. (www.photoimagenews.com/instant.htm)

During the ongoing MUFON investigation, Walters was provided with a Nimslo stereo camera, consisting of four independent fixed lenses. This stereo camera was used specifically as a control to help rule out the possibility of a perpetrated hoax. The Nimslo camera was chosen specifically for a variety of reasons. These include the fact that, by design, one feature unique to this particular camera was double exposure prevention, thereby ruling out hoaxing by this method. [42]

Chain of evidence was further established by virtue of the fact that the camera entered Mr. Walters's possession loaded with film and the film compartment had been sealed in wax, thereby providing field investigators with a means to establish evidence of tampering.

November of 2007 marked the twentieth anniversary of the initial event that would spawn the Gulf Breeze UFO wave. The technologies available to MUFON and their cadre of field investigators are infinitely more advanced than those available when the case was

Stereo camera, similar to the one used in the Gulf Breeze investigation

originally investigated. Perhaps re-examining the case some twenty years after the fact will yield discoveries previously unavailable due to lack of resources.

Photographic Analysis of the Gulf Breeze Photographs

It is the opinion of this writer that the events that occurred at Gulf Breeze, Florida from 1987 until 1990 may well be one of the better cases in the history of modern ufology if, in fact, they are genuine. The Gulf Breeze UFO case reports went on to become a hot button issue in the field of ufology, dividing much of the research community between those who felt that the photographs were likely genuine, while others expressed serious reservations as to their validity.

Dr. Bruce Maccabee, optical physicist and noted UFO researcher, notes in his April, 1989 article on this case for the *MUFON UFO*

Journal that he and other researchers working on the photographic components of the case *"diligently searched for irregularities and oddities which would clearly prove that hoax techniques were used in creating the pictures"* and found none.

Later, Maccabee went on to state – quite correctly – that the entire weight of any given UFO investigation which involves photographs cannot be based solely on studying the images because photo evidence is unfairly biased towards a hoax determination.

To this end, Dr. Maccabee writes:

> *"This bias comes about because a single UFO photo could contain evidence that, by itself, proves the sighting is phony. For example, there might be a vertical linear image above the image of a "UFO" indicating that it was suspended by a string or thread; there might be an image below the UFO image suggesting that the UFO was supported from below; there might be a discontinuous variation in brightness or color between the edge of the UFO image and the background indicating a photo montage; or there might be overlapping images indicating a double exposure."*

Gulf Breeze – genuine, hoax, or inconclusive?

There are some in UFO research field who believe that during the course of his investigation, Dr. Bruce Maccabee, an optical physicist employed by the United States military, lost his objectivity and certified obvious hoax photographs as genuine. This supposition is not only insulting to a researcher of Dr. Maccabee's caliber, but it presupposes that Maccabee would stake his military career and professional reputation on the certification of photographs that might easily, given the passage of time and advancement of technology, be easily proven false.

The argument in the UFO field that *"extraordinary claims require extraordinary evidence"* is put to the test here. Those who might say that a career scientist such as Dr. Maccabee forgot his scientific objectivity are, in my experience, doing just that – making an extraordinary claim.

Maccabee's photo analytical experience neither began nor ended with the Gulf Breeze case. From that day to this, he has served with distinction as a photo and video analyst for the Mutual UFO Network.   From January, 1988 through January, 2007, Dr. Maccabee wrote and presented papers at no fewer than seven international symposia sponsored by the Mutual UFO Network. He has also published countless articles dealing with photo analysis and other specialized investigation in scholarly journals.

There are arguments to be made for each of the three possible conclusions to the Gulf Breeze case. There is compelling evidence to suggest that a high percentage of the photographs taken by Edward Walters over a period of approximately two and half years are, in fact, genuine. The sheer volume of photographs attributed to Walters, coupled with the fact that images were captured on multiple cameras, across multiple photographic platforms (i.e. traditional 35mm, instant Polaroid film, and multiple exposure stereo photography) suggests that perpetrating a hoax is highly unlikely. It is possible that Walters may have staged a percentage of the Gulf Breeze UFO photos, but even if all the Walters photographs were excluded from the Gulf Breeze case file, there would still be a number of case reports from individuals other than Walters, as well as cases that fall well outside the generally accepted timeline of reports attributable to Walters.

This being said, the fact remains that a model resembling one of the anomalous craft photographed by Walters was found in a home where Walters once lived. While this model does not match the photographed object in all respects, it is, in fact, similar enough to the object seen in several of the Gulf Breeze photographs so as to raise the question of forgery.

This writer is skeptical of the model explanation, discussed at length in an article by longtime UFO researcher J. Antonio Huneeus in the August, 1990 *MUFON UFO Journal*, as the source of all of the anomalous photographs attributed to Walters for various reasons. Most notably, the odds of a single individual being a skilled photographic forger across the multiple photographic platforms utilized during the course of the Gulf Breeze investigation, particularly prior to the advent of digital photography and digital imaging software, would almost certainly preclude the hoax hypothesis from being viable.

As with any matter of science, there is the possibility of simple human error. This scenario, as it relates to questionable photographs among the Gulf Breeze case file, cannot be dismissed outright, but there is no definitive evidence to suggest any misstatements with regard to the photo analysis conducted by Dr. Bruce Maccabee on behalf of the Mutual UFO Network.

Regardless of whether or not the Gulf Breeze photos are ever definitively determined to be genuine anomalies, elaborate hoaxes, or something in-between, the Gulf Breeze events are still very much relevant in the field of UFO research. Thanks in part to the publication of at least two books on the case, as well as the advent of the Internet, the Gulf Breeze photographs and the controversy surrounding them are still furthering the debate some twenty years after the initial case report.

While a number of photographs covering the 1987 through 1990 timeframe show the earmarks of possible forgery, an equal or greater number from the same period were shot under conditions that preclude them being attributed to a hoax.

The stereo photographs are of particular interest in that they were taken under scientifically controlled conditions, which rules out the possibility of a hoax in that particular instance.

Even if the 35mm and Polaroid photographs attributed to Edward Walters are discounted as the product of forgery, the stereo photographs as well as the photographs and video footage shot by other individuals (including MUFON personnel) in Gulf Breeze and the surrounding area need to be adequately explained.

While Edward Walters has become synonymous with the events of Gulf Breeze, several people have photographed objects in the sky over the Gulf Breeze/Pensacola area. It is important to note that a number of these photographs are taken outside of what has become the established timeline (1987 through 1990) for the Gulf Breeze UFO wave. Specifically, a number of individuals have reported or photographed anomalous aircraft over Pensacola over the weekend of July 6-8, 1990.

Ironically, this is the weekend during which the Mutual UFO Network hosted its international symposium, with Pensacola as the host city.

It is interesting to note that, even as the 1990s wore on, objects similar to those reported by Edward Walters continued to be photographed in the Pensacola area by several witnesses including the current MUFON state director for Florida, G. Bland Pugh, who captured an object very similar to a Walters photograph as late as 1996.[43]

While the UFO incidents occurring in the late 1990s should be viewed as separate incidents from the 1987-1990 wave at Gulf Breeze, there is a direct connection between the two. Regardless of whether or not some or all of the Walters photographs taken from November 11, 1987 through May 1, 1988 are the product of hoaxing, the fact remains that UFO sightings continued to be reported in and around Gulf Breeze by residents through the first quarter of 1990.

The second wave of UFO sightings, if it can be called that, to hit the Gulf Breeze area could be said to last from 1991 until at least 1996. On at least two occasions during this period, in 1993 and 1996, respectively, images resembling some of the Walters photographs were shot in the Pensacola/Gulf Breeze area. These photographs, in and of themselves, are significant in that they were shot long after the original Walters photos and that, having been photographed by a photographer other than Walters, the possibility of a hoax can initially be ruled out.

Both photographs to which I am referring were photographed by a staff member of MUFON. The earliest of these photos appeared some six years after Walters's initial report in November, 1987 and, thus, could not be directly related to the earlier Walters photographs.

Following the simplest of scientific protocols, Occam's Razor, the most logical assumption likely being the correct one, even if all of the photographs shot by Edward Walters from the period of November, 1987 through May, 1988 are discounted as forgeries (I believe they should not be), that still leaves the 1991 through 1996 events to be explained.

Some of Edward Walters's photographs could be forgeries. It is possible. Could dozens of photographs, taken over a period of six months, using multiple cameras and utilizing multiple photographic formats all be forgeries as well? The odds of all of those photographs being the subject of fraud is, pardon the pun, astronomical.

As to the second series of events at Gulf Breeze, with multiple witnesses being involved over a period of years, the opportunity for fraud does exist, but by the same token, the similarity of some of the later photographs, particularly the 1996 photograph, to earlier Walters photos is certainly striking. One case seemingly confirms the other.

It is my personal and professional opinion that, while some early photographs from Gulf Breeze exhibit characteristics that suggest a possible forgery, taken in concert with later photographs, shot some six to nine years after the fact, it would seem that the logical determination in this case would be inconclusive, with a trend towards the *"genuine event"* hypothesis in the later cases.

The Gulf Breeze photographs in historical context

Setting aside all of the conflicting theories, points of view, and arguments on both sides of this complex case, the fact remains that, regardless of the final determination as to the validity of the 1987-1990 UFO photographs shot by Edward Walters over Gulf Breeze, Florida, there was a second series of events, which did not center around Walters's testimony, that continued in the Gulf Breeze area, producing additional photographic images, throughout most of the decade of the 1990s. Two series of separate, distinct Gulf Breeze UFO events continued, with documentary investigation by researchers from MUFON, for a period of nearly a decade. For this reason alone, the Gulf Breeze events are historically significant within the field of Ufology, regardless of the questionable origin of some of the accompanying photographs.

Regardless of anyone's personal opinions with regard to the Gulf Breeze matter, the problem I see with this specific case is that most members of the general public, as well as a percentage of UFO researchers, do not make what I feel is a vital distinction between the earlier events, tied almost exclusively to Walters, and the later events in the Gulf Breeze/Pensacola area, with an additional witness pool, additional evidence, and case data. Viewing both waves as a continuous series of events does a disservice to the research itself. If this distinction is not made, the data from the Walters reports runs the risk of merging, at least in the minds of a few, with the later cases, running the risk of prejudicing the outcome of the investigative process.

Whether or not Edward Walters engaged in the most fantastic photographic UFO fakery since Eduard "Billy" Meier is immaterial. The Gulf Breeze UFO incidents were, perhaps, the first UFO sighting wave to be documented to such a great extent, utilizing multiple photographic formats over such an extended period of time. It is, conversely, perhaps the last significant photographic UFO cases to be documented exclusively using analog film technology. With the shift to digital technology occurring in the latter half of the 1990s, the analysis of photographic images only became more complex. The gray area of ufology, the ability to discern the validity of questionable images, would only get hazier moving forward.

If one only looks at the latter Gulf Breeze cases, investigated as late as October, 1996, the case must remain open. At the very least, Gulf Breeze remains an intriguing point in the history of Ufology. If it serves no additional purpose than to further debate, that is reason enough for the events at Gulf Breeze to occupy a prominent place in the history of modern ufology.

(Endnotes)

[39] http://en.wikipedia.org/wiki/Gulf_Breeze_Florida

[40] www.cityofgulfbreeze.com/events.html

[41] Walters Ed, and Walters, Frances. The Gulf Breeze Sightings. New York: Wm. Morrow & Co. Inc., 1990.

[42] www.stereoscopy.com/cameras/nimslo.html

[43] www.ufocasebook.com/gulfbreeze.html

15. The STS-48 Video Footage

The 1991 flight of the Space Shuttle *Discovery* known as STS-48 has become perhaps one of the better known cases in modern ufology due to the fact that it apparently offers researchers something previously unheard of – documented, verifiable footage of an unidentified object exhibiting characteristics which seem to contradict conventional physics.

In a very real sense, the STS-48 event represents the proverbial "smoking gun": a bona fide UFO, regardless of what definition you ascribe to that term.

On the evening of Thursday, September 12, 1991, the Space Shuttle *Discovery* was launched from the Kennedy Space Center at 7:11 PM Eastern Time. The Shuttle's primary payload on the planned six-day mission was the deployment of the Upper Atmosphere Research Satellite (UARS) on September 15, the third day of the mission.

Interestingly, among the planned experiments slated for that weekend included the testing of photographic equipment, specifically a Nikon F4 35mm camera that had been converted to produce digital images to be downlinked directly to the Electronic Still Camera Laboratory at the Johnson Space Center in Houston, Texas.

According to NASA's own press kit on the STS-48 mission, the camera equipment onboard *Discovery* included the F4 camera, which, after conversion to a digital format, had a resolution capacity of 1 megapixel, or one million pixels. *Discovery* also carried several lenses, at a variety of focal lengths, including 20mm, 35-70 mm, 50mm, and 180mm. While substantial for use in documenting the results of onboard experiments, these camera components would not possess the needed resolution to document objects seen any great distance from the shuttle.

It is important to note that while the camera equipment carried onboard *Discovery* at the time may sound antiquated by today's standards, digital photography technology was still in the developmental stages in 1991, and consequently, the photographic technology onboard was significantly more advanced that what would have been available to the general public, even though the onboard cameras and equipment were specially-retrofitted versions of commercially available products.

It is interesting that the official plan for the mission included photographic testing; as the five member crew of *Discovery* would go on to capture what is, in this writer's opinion, some of the finest photographic evidence to date of what seems to be an object clearly demonstrating capabilities beyond conventional physics as we presently understand it.

The STS-48 video footage as documentary evidence

The STS-48 video footage, which has become the central element to this case, is of immense importance due to the circumstances under which it was first captured. By virtue of having been photographed by video cameras onboard the Space Shuttle *Discovery*, a provenance, known in law enforcement and legal fields as 'chain of custody' was immediately established.

Due to the fact that the video cameras in question were on the exterior of the shuttle, the possibility of a hoax can be dismissed immediately. The remaining available options are equally intriguing, regardless of which viewpoint you take on the subject. Discounting the possibility of a hoax, the options remaining under consideration are that the questioned object is either a misidentified terrestrial object or artifact of spaceflight, such as ice crystals or "space junk", as NASA argues, or something much more intriguing altogether – a true UFO – an unidentified flying object.

It is my personal opinion, as well as the opinion of numerous others within the field of ufology that the STS-48 video footage shows an object that does not possess characteristics of any object that might otherwise be mistaken for a UFO. My personal feeling is that, regardless of what the object in the video may have been, it most certainly fits the broadest definition of the term UFO, in that it was, without question, a physical object and not a lens flare or photographic anomaly. It has not, to the best of my knowledge as of this writing, been definitively identified, and was operating under what can reasonably be defined as intelligent or independent control in open space.

For these reasons, I believe the video taken by the crew of STS-48 on September 15, 1991, from 20:30 to 20:45 Greenwich Mean Time, deserves serious scrutiny from the scientific and photographic communities, in hopes that some consensus may be reached. I believe that the jury is still out when it comes to the question of what is shown in the questioned video. Until this question can be satisfactorily resolved, I believe that it is incumbent upon science to take a serious look into this footage, as it may hold answers to questions not only in the field of physics, but numerous other disciplines as well.

The STS-48 video footage compared to other photo evidence cases

The STS-48 video footage, when compared to other photographic evidence UFO cases, clearly stands out. It is unique in that this case involves video footage, as opposed to still images. The difference between this case and others, whose primary evidence is still photography, is that the STS-48 video footage provides a unique frame of reference for investigators which would otherwise be unavailable with still photography – the documented movement of the object as opposed to a freeze frame during the encounter.

The fact that the STS-48 footage was coming by way of a direct feed from the Space Shuttle *Discovery* is also, in my view, an important factor to consider in that there was no way for the footage to become corrupted or otherwise altered between the moment of capture and the moment of reception, whether at the Kennedy or Johnson Space Centers, or subsequent broadcast over the airwaves.

As evidenced by the research of physicist Dr. Jack Kasher of the University of Nebraska-Omaha, the STS-48 video holds significant information regarding the questioned object.

For these reasons, among others, the STS-48 video compares quite favorably to some of the more well-known photographic UFO cases, such as the McMinnville, Oregon and Santa Ana, California events discussed previously.

It is my personal opinion that, with the further passage of time and perhaps renewed research, the STS-48 video footage might be seen as perhaps one of the more important UFO incidents of the late twentieth century, as well as a significant piece of evidence that tips the scales of science and academia towards accepting as fact the existence of life beyond the constraints of this planet.

The STS-48 incident as the tipping point for NASA

The STS-48 incident can be said to be "the tipping point" for NASA in the respect that the live broadcasts of shuttle missions via NASA Select suddenly stopped following the STS-48 mission.

Whether or not there is a cause-effect relationship between the two, the fact remains that this is, in fact, the conclusion numerous members of the UFO research community have come to over the years.

Personally, I can't say definitively that the broadcast video of September 15, 1991 was the direct cause of NASA Select being taken off the air; however the two events certainly do appear to be related.

I may not be able to make a definitive statement as to the connection between the STS-48 mission and what appears to be an almost immediate policy change by NASA, however, in my opinion, there appears to be no "alternative theory of events", to borrow a legal term.

This series of events, in my view, is one of the primary factors which make this case so very interesting from a research standpoint. A logical person could easily come to the conclusion, after having seen the video footage in question and make a very viable argument for increased funding for NASA specifically and the sciences in general. However, NASA apparently took an alternate position, dismissing the footage aired on its own cable channel as well as the entire event.

It is not my intention to attempt to sway the debate on this issue in one direction or the other; instead, I feel it is most prudent to state my own case and allow others to formulate their own opinions based on the material presented.

The Space Shuttle Discovery
Courtesy NASA

Technical analysis of the STS-48 video footage

To date, the most in-depth technical analysis of the STS-48 video footage has been conducted by Dr. Jack Kasher, formerly of the Lawrence Livermore National Laboratory and professor emeritus of physics at the University of Nebraska at Omaha.

In addition, Dr. Kasher is a well-known member of the UFO research community, having served for many years as both State Director for Nebraska and consultant in Physics to the Mutual UFO Network.

In the interest of a full and complete recounting of events, I feel it is important to delineate the chain of custody, with regard to the original video footage, which was recorded from the NASA Select cable television channel by Maryland UFO researcher Donald Ratsch. Ratch's initial contacts included Vincent DiPietro of the NASA Goddard Space Flight Center in Greenbelt, Maryland, as well as Congresswomen Helen Bentley. Subsequently, DiPietro found the footage to be so compelling that he was also inclined to contact his representative, Congresswoman Beverly Byron, also of Maryland.

As a result of the DiPietro inquiry, Congresswoman Byron forwarded the video to Congressman George Brown, Jr., then the chair of the Congressional Committee on Science, Space and Technology, who proceeded to show the video to several members of his staff, presumably to elicit their opinions on the matter. Their conclusion was that the glowing objects seen in the video were likely ice particles that had been ejected as liquid water by *Discovery*, freezing in the sub-zero temperatures of space. The consensus opinion, as to the sudden change in direction, was attributed to the Shuttle's own attitude adjustor rockets, which are used to change the Shuttle's orientation while in orbit over the Earth.

Subsequently, the video was passed on by Congresswoman Bentley of Maryland to Martin Kress, an assistant administrator for legal affairs at NASA, who also made the video available to no fewer than four NASA scientists, who concurred with the earlier opinion that the objects seen on the videotape were likely ice particles.

It was at this point that Kasher began his own research, picking up where Ratsch's own cursory research left off. Perhaps the most vital question – on which the entire case hinges – is whether or not the objects seen passing into the field of view of *Discovery's* cameras were, in fact, ice particles.

In the interests of brevity and clarity, I will rely quite heavily of Dr. Kasher's own words to best explain his research to a mass audience, as the highly technical nature of this case makes a thoroughly detailed explanation prohibitively difficult.

To this end, in his paper published by the Fund for UFO Research, Dr. Kasher writes:

> *"To the best of my knowledge, the individuals who suspected ice particles merely watched the videotape, and did no further scientific analysis. Also, it appears from the wording of the letters that these individuals did not comment specifically about the two streaks that went through the picture several seconds after the glowing objects accelerated. It is possible that they meant to include the streaks with the other objects when they spoke of ice particles."*

Dr. Kasher further states his reasons for disagreeing with the ice particle hypothesis by saying,

> *"My analysis is a study of the actual motions of the objects as observed on the videotape, and the inconsistency of their behavior with what would be expected of ice particles. As such, it does not depend on whether the vernier rockets actually fired or not. As a matter of fact, if no verniers fired during this time period, my analysis would be unnecessary – it would not be possible for the objects to be ice particles accelerated by the vernier adjustor rockets."*

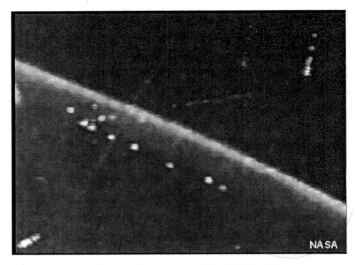

Time-lapse video capture of the STS-48 video
Courtesy NASA

Perhaps the most basic point made by Dr. Kasher, somewhat crudely illustrated in the video capture from STS-48, is the fact that purported ice particle seen in the video slows down, appears to stop, then radically changes course, accelerating rapidly.

Even taking into consideration a lack of gravity, Kasher astutely notes, *"It is difficult to conceive of a mechanism by which a rocket exhaust could stop an ice particle and allow it to sit at its location for about half a second, then accelerate back to the right."*

In the most basic of terms, it was Kasher's professional opinion as a physicist that the glowing objects present in the STS-48 video behaved in a way that was inconsistent with ice particles in space, even when one took into account the possibility of the shuttle's own adjustor rockets factoring into the equation. In short, there had to be an explanation other than NASA's official explanation of events.

Dr. Jack Kasher was instrumental in the analysis of the STS-48 video feed.
Courtesy MUFON Nebraska

The STS-48 Incident in historical context

The STS-48 incident is, in the opinion of this writer, terribly significant when one looks at the recent history of the UFO phenomenon. It is this case that provides a true proverbial smoking gun to the UFO research field in that the provenance of the video footage is indisputable.

Dr. Jack Kasher is to be commended for his exceedingly thorough study of this video. Provided that Kasher's research stands the test of time moving forward, the STS-48 event could very well be viewed by future generations of researchers as one of the seminal pieces of evidence to come out of this era.

This case is also exceedingly important, in my view, given that it is not a photographic case, but a video case. This subtle distinction is nonetheless significant in that this case quite literally unfolded in real time, whereas the sequence of events in many photo cases is much less clear, given that the best data available, with regard to time is often an approximation.

As stated previously, it is my belief that with the further passing of time, the value of both the footage itself as well as the research conducted into this case will be reevaluated and, I believe, judged to be perhaps one of the best evidence cases of the modern era.

16. The Phoenix Lights

On the evening of March 13, 1997, the city of Phoenix, along with most of the state of Arizona and parts of Nevada, experienced a series of spectacular light displays which were witnessed multiple times by multiple citizens. Among the witnesses were commercial airline pilots, architects, police officers, air traffic controllers, medical professionals, and others. At least one airman from Luke Air Force Base was also among the witnesses. Many of these witnesses captured amazing photographs and video tape recordings of the lights.

The lights were seen not only in and around Phoenix, but throughout the State of Arizona over a period of three hours, between 7:30 and 10:30 PM Mountain Time, covering more than three hundred miles from the Nevada state line, through Phoenix, and on to the northern edge of Tucson.

Obviously, with such a lengthy duration, reports of mysterious lights in the night sky were not exclusive to the City of Phoenix.

The first report came from Henderson, Nevada, from a former police officer and his family. Then the lights were seen over Paulden, Arizona. Following their visit to Paulden, the lights were seen in Prescott Valley, a distance of thirty miles to the south, which the lights traveled in the span of just one minute. The next sighting was over the town of Dewey, Arizona, some ten miles to the south of Prescott Valley. Additional reports came in from Chino Valley, Tempe and Glendale. Then, the lights arrived in Phoenix. It was from here that many of the photographs and videos were taken. Following their visit to Phoenix, the lights continued on their leisurely journey in a southeast direction towards Tucson and finally, Kingman, Arizona, where the last sightings of the anomalous object were reported.

The Phoenix Lights as documentary evidence

The Phoenix Lights are unique in modern ufology, in the opinion of this writer, in that it was not only a mass sighting event of the highest order, but also perhaps the clearest example of documentary evidence of the UFO phenomenon in recent memory, if not in the history of the phenomenon, given that first-hand witnesses captured the event on film, video, and other media from a wide variety of locations not only in the City of Phoenix, but throughout the State of Arizona.

The Phoenix Lights represents, in my view, perhaps "a perfect storm" – a perfect confluence of circumstances which would allow the case to become one of the most well-known events in the history of the phenomenon.

In the interest of clarity, I will dissect the event point by point as I state my own case for the Phoenix Lights event.

As mentioned previously, the events of March 13, 1997 were not confined solely to the City of Phoenix. If that were the case, there would be significantly fewer case reports from which to glean information, which could, in theory, skew the data available to researchers.

The Phoenix Lights case also makes a case for the value of first-hand witnesses. The thousands of witnesses in the city of Phoenix that night ultimately produced photographic and video evidence from virtually every conceivable angle over a period of hours. The photo and video evidence is truly the backbone of this case, providing researchers with images captured under a host of conditions, allowing these professionals to conduct analyses based on virtually any conceivable variable.

Perhaps the simplest evidence, with regard to the Phoenix Lights, of the value of documentary evidence is the fact that the city of Phoenix is home to over 1.5 million people. In a city the size of Phoenix, the likelihood of being among the witnesses that night was incredibly high, simply given the population.

With 1.5 million residents, it is safe to say that the material presently in the public domain is likely only a fraction of the documentary evidence in existence.

Of particular importance, in my view, is the video footage. While still images offer a trained professional a wealth of information, the video evidence brings another dimension to the research. Video images allow experts to see events unfold in real time.

Phoenix, Arizona skyline, 2001
Courtesy Jon Sullivan

The Phoenix Lights, mainstream media, and military

The extraordinary events in Nevada and Arizona, on the night of March 13, 1997, received little or no mainstream press coverage at the time, something all too familiar to UFO researchers. It was not until several months later, and then on subsequent anniversary dates, that mainstream media coverage occurred, in media outlets including CNN and the *Arizona Republic* newspaper.

However, much has since been written on the topic of the Phoenix Lights and is available in copious detail via the Internet.

It is not my intention to re-hash reasons for the lack of press coverage of UFO sightings by the mainstream media. This point, in my view, is immaterial. It is far more interesting to examine the opposing viewpoint – the impact of the event on the paradigm of UFO research. It is my feeling that the Phoenix Lights incident of 1997 represented "critical mass", for want of a better term, with regard to the media's treatment of the UFO subject.

I believe that looking back on the history of Ufology in the near term, even going back to the relatively recent past, a reasonable argument can be made that the UFO subject is viewed today far more sympathetically and as a matter of science thanks, at least in part, to the photo and videographic evidence that came to light as a result of the Phoenix Lights sightings of March, 1997.

Prominent ufologist Dr. Bruce Maccabee has referred to the Phoenix Lights as one of the most important UFO events of the last fifty years, and yet it failed, at the time of the event, to garner major mainstream media coverage. It is curious how an event such as the Phoenix event could escape media attention, regardless of the explanation – terrestrial or otherwise.

Central to the case of the Phoenix Lights is the involvement, as in many other UFO case reports, of the United States military – in this case, the personnel of Luke Air Force Base.

As witnesses on the ground throughout Phoenix were reporting the lights, witnesses in the air were seeing the same thing. The lights (object) passed over Phoenix's Sky Harbor Airport, where they were seen by air traffic controllers who did not the object or objects on radar. At least one commercial airline pilot reported the lights in the air, asking Sky Harbor tower what "these nine lights were".

As is typical in these types of unusual airborne events, conflicting information was received. Many witnesses called Luke Air Force Base with their reports, the operator telling them they were being flooded with calls. However, just a few days later, Luke Air Force Base officials publicly stated they had received no calls regarding the incident.

The Air Force's official explanation was further compromised when a local UFO researcher was able to produce telephone records clearly showing that he himself had called several times that evening. Also, several witnesses had received the telephone number to the National UFO Reporting Center (NUFORC), based in Washington State, from personnel stationed at Luke Air Force Base.

One witness who called the NUFORC hotline claimed to be an airman at Luke Air Force Base and reported that the Air Force scrambled two F-15 fighters and that one of the fighters had intercepted the object over Phoenix. A cursory search of the F-15 inventory at Luke was inconclusive; however the airman's account was confirmed by a long haul truck driver who reportedly watched the UFO intently as he was driving down Interstate 17. The truck driver claimed the fighters were F-16s and that there were three which blasted out of the base with afterburners on full. He added that the aircraft headed straight for the UFO which then shot upwards and disappeared instantly as the fighters approached.

This story was corroborated when a UFO researcher submitted a Freedom of Information Act (FOIA) request to the Air Force for information which confirmed that F-16s were sent out that night.

Not surprisingly, the Air Force's official position on this incident is that the jets involved were on a routine training mission and had no involvement with the reported lights.

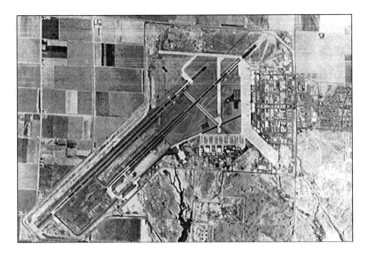

Luke AFB, Arizona
Courtesy US Air Force

The military's statement that the flight of military aircraft over the city of Phoenix was unrelated to the anomalous lights reported on the evening of March 13, 1997 is, in the view of this writer, suspect at the very least given that the Air Force initially denied any knowledge of the lights, then released an official statement – the lights reported by hundreds of eyewitnesses that evening were nothing more than flares.

The military claimed that the flares had been released by an Air Force A-10 over the Gila Bend Bombing Range, which is located southwest of Phoenix on the other side of the mountain range from where the sightings occurred. The Air Force explained that the A-10s were on a special training mission from David-Monthan Air Force Base located in Tucson and that they had released the flares at about six thousand feet, ignited at approximately three thousand feet and had burned out by five hundred feet altitude.

Regulations do not allow aircraft to land with flares on board.

One can only speculate as to the military's motivation for the abrupt about-face regarding the aircraft involved in the Phoenix events; however it is my opinion that the flare explanation is more convenient and palatable from a public relations standpoint than

attempting to explain the appearance of an anomalous object or objects over a densely populated major American city.

A closer examination shows that the timeframe the Air Force provided for the flares having been dropped was some 42 minutes after the reported sightings in Phoenix.

Flares may have been a reasonable explanation for the lights seen over Phoenix, however fail to explain the sightings over other areas of Arizona, as well as in the State of Nevada. Flares simply cannot travel over such a large area.

Additionally, several witnesses reported the object they had seen as being a distinctly solid object. For obvious reasons, flares must be excluded as the cause of these particular sightings.

The following abstract, coming from the National UFO Reporting Center, further argues against flares as the exclusive cause of the Phoenix incident:

The object was capable of very rapid flight, probably supersonic, although few witnesses reported any noise. The object was reported heading generally to the southeast over Henderson, Nevada at 18:55 (Pacific Time) and was next reported heading to the south near Paulden, Arizona 22 minutes later. Within approximately one minute of the Paulden sighting, the object was in the vicinity of Prescott Valley, Arizona, roughly 30 miles to the south. The object then appeared over Phoenix where reports have it hovering for 4 – 5 minutes in the vicinity of the intersection of Indian School Road and 7th Avenue.

Clearly, what is being described here exhibits characteristics inconsistent with military flares.

Technical analysis of the Phoenix Lights

The Phoenix Lights case is unique among photo and videographic evidence cases, in my view, as it was one example of a case in the fairly recent past where computer technology and enhancement aided researchers in the analysis of both photo and video material.

Those who have read *The Phoenix Lights* by Lynne D. Kitei, M.D., will be familiar with the name Jim Dilettoso. Mr. Dilettoso is a noted photographic analyst who conducted an extensive analysis of the photographs taken by Dr. Kitei. Dilettoso had been involved with the Aerial Phenomenon Research Organization (APRO) in the 1970s and had been instrumental in developing a procedure to test questioned photographs that also adhered to the scientific method.

Mr. Dilettoso was the first person to publish a paper on how to test UFO pictures scientifically, whereas other researchers, such as Dr. Bruce Maccabee and Dr. Richard Haines applied existing protocols to photographic analysis of unknown objects.

Before one can understand the detailed analysis inherent in any one photographic evidence UFO case, it is easier to understand a broad hypothetical scenario, before injecting the actual facts of the case.

Using the Phoenix events as just such an example, it is important to consider that the most common piece of evidence, aside from a direct witness statement may well be a photograph. This will be the starting point of the overview.

When conducting an analysis, it is vitally important to obtain, either directly or indirectly, the highest resolution photographs possible, as Jim Dilettoso did in this case.

Next, a specialist (such as Dilettoso, Maccabee, Haines, or others) would proceed to input the questioned image into a computer in order to study various attributes of the image, such as levels of light and shadow, image density, and other features. While a criminal defense attorney would argue that this type of testing constitutes manipulating an image so as to make it unreliable, from a legal standpoint, it is important to realize that the computer software now used in the analysis of questioned photographs serves the same role and function as equipment found in photographic darkrooms as recently as the past decade.

Simply put, the advancement in digital technology allows photographic experts to work smarter, not necessarily harder.

Using software, trained professionals can differentiate between different forms of light, such as streetlights, aircraft lights, flares, etc. Each of these types of light produce a specific "signature" when analyzed and filtered according to the visible spectrum. Simply put, by comparing the "signature" of an as-yet unidentified source to that which has been previously identified, a researcher can determine, to a reasonable degree of certainty, whether or not a questioned light or object can be readily identified.

Analysis of the Kitei photographs, for example, showed the photos to be unlike anything Dilettoso had previously encountered in his years as of photographic experience. There was no match between the unknown and known reference samples which he kept on file. He could, however, tell that the orbs were light emitting and not light reflecting, meaning that they did not light anything around them; not the clouds or the ground. Also, the light was diffuse, meaning the light was inside the object.

Dilettoso also described the light as being a very pure, precise light and that while the amber orbs were bright, it did not appear that their purpose was illumination.

After in-depth analysis by a qualified professional with knowledge of photography and optics, at least one set of photographs stemming from the 1997 Phoenix Lights event seems to be of unknown origin.

The Phoenix Lights as a 'special circumstance' case

The Phoenix Lights can be viewed as a "special circumstance" case for several reasons, as there are several factors which set it apart even from similar cases within the field of ufology.

Firstly, this incident took place over a major population center in the United States, lasted several hours, was documented and reported by hundreds of eyewitnesses, while undoubtedly being witnessed by tens of thousands of people.

Secondly, the sheer duration of this incident clearly ranks as one of the longest single events in the history of ufology, at least to the knowledge of this writer.

Thirdly, the military and aviation component of this incident is vitally important, as it offers yet another avenue of investigation to researchers, potentially including radar data from both Luke Air Force Base as well as commercial flights in and around the area at the time of the incident.

For these reasons, among others, the Phoenix Lights incident certainly deserves mention as a case which stands out from other UFO field reports that involve photographic components.

The Phoenix Lights in historical context

Historically, UFO photographs have tended to be distant, fuzzy objects or unknown lights. Conversely, we have sketches made by single witnesses or small groups.

With regard to the Phoenix Lights, we have numerous clear photographs and solid videos taken by different witnesses throughout the city, all showing the same object.

In this case, the object seen over the City of Phoenix operates in such a way that the question of intelligent control, whether human or otherwise, seems a foregone conclusion.

Much like the 1942 incident over Los Angeles, California, this object appeared to operate not only under intelligent control, but also with a certain level of disregard for onlookers.

The Phoenix case is significant, if for no other reason than being the largest mass UFO sighting of the modern era. The notion that an incident such as this could occur over a major metropolitan area in front of possibly more than 1 million witnesses is unprecedented.

Also unprecedented is the research into the photo and videographic evidence captured by the many witnesses that evening. It is also incredibly significant to note that, upon analysis, at least some of the lights photographed at night were determined to be, in one researcher's professional opinion, a bona fide unknown.

What exactly was seen that night over virtually the whole of the State of Arizona remains unclear. In the opinion of this writer, the exact origins, intent, and nature of the object sighted that night are, to a degree, immaterial. The Phoenix Lights were legitimate UFOs in the most basic definition of the term. The object or objects were both unconventional and unidentified, were flying, and clearly had mass. Whether or not this case is ever definitively resolved, it is the opinion of this writer that this incident clearly warrants inclusion alongside the most strenuously researched cases in the history of the UFO phenomenon.

17. The Millstadt, IL Flying Triangles

One of the more interesting UFO events in recent memory occurred just five days into the twenty-first century as anomalous objects were sighted in no fewer than four communities in southern Illinois, as well as parts of the State of Missouri.

At approximately 4:00 AM on the morning of January 5, 2000, a long-haul truck driver in Highland, Illinois, some two dozen miles from Saint Louis, Missouri, would make the first of what would be several reports of an anomalous dark triangular or delta-shaped flying craft. Subsequent reports, particularly germane to this account, would be reported between 4:10 and 4:15 AM by a police officer in Lebanon, Illinois, as well as by an additional officer in the neighboring jurisdiction of Dupo, Illinois at 4:29 AM.

While the incident encompassed multiple police jurisdictions in southern Illinois, for the sake of both brevity and clarity, I will refer to the incident exclusively as the Millstadt, Illinois event, in keeping with the most common citation of the case in the literature.

This event is perhaps most interesting due to the fact that not only were multiple witnesses involved, but also the fact that several of the parties involved were law officers, a group many in the field consider to be "professional observers", for want of a better term.

Photographic evidence of triangle-shaped anomalous craft worldwide

The Millstadt, Illinois events were far from unique in their reports of dark, triangular-shaped craft. The fact is that these craft have been seen repeatedly throughout the world dating back a minimum of two decades.

Since at least 1989, reports of black (or dark colored) delta-shaped aircraft have been common throughout the world, most notably the well known wave of reports over Belgium in the late 1980s and early 1990s. In addition to the sightings in Belgium, similar craft have also been reported in Israel, Canada, and numerous locations in the United States.

The Millstadt, Illinois event is perhaps the latest and best known in this string of events.

While photographic evidence exists of the Millstadt sighting of January 5, 2000, the only readily available photograph, so far as this writer is aware, is a single Polaroid shot by Dupo, Illinois police officer Craig Stevens at approximately 4:29 AM, enhancement of which for purposes of publication would have effectively washed out the visible portion of the photo, rendering it useless for purposes of comparison with earlier documented triangular craft seen in other parts of the world.

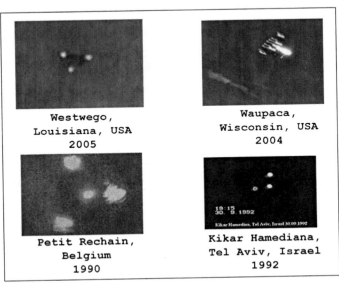

| Westwego, Louisiana, USA 2005 | Waupaca, Wisconsin, USA 2004 |
| Petit Rechain, Belgium 1990 | Kikar Hamediana, Tel Aviv, Israel 1992 |

Various triangular craft have been reported over the years, dating back to the late 1980s.
Courtesy the Author

As evidenced by earlier case reports, the dark triangular or delta-shaped craft reported in the vicinity of Millstadt, Illinois seems to be part of an ongoing trend and even without photographic confirmation of the event, due to Officer Stevens's malfunctioning Polaroid, the reports made by the supporting witnesses allow for a fairly clear, detailed narrative of the event to take shape.

The role of law enforcement and first responders as sources of evidence in the Millstadt event

The Millstadt event is, in my view, a very clear example of the value of law enforcement officers as witnesses and sources of evidence in a UFO event. The credibility of the Millstadt event is established, it can be argued – even without detailed photographic evidence – thanks to the detailed sketches provided by both police witnesses, Dupo officer Stevens and Lebanon officer Barton.

Not to discount the civilian witness testimony in this case, but the testimony of officers Stevens and Barton is particularly important. Absent the photograph taken by officer Stevens, the secondary reports made by the officers truly provide the backbone of the Millstadt case.

The officers' testimony is particularly important, as in other events involving law enforcement officers, given that police officers are trained observers and as such, are, to an extent, considered to be more reliable than a civilian witness.

Stating that a police or law enforcement witness is considered to be more reliable than a civilian witness is not to say that a civilian is *unreliable,* simply that members of law enforcement are, by their nature, more skilled at making note of minor details, which might otherwise escape a civilian witness.

This is particularly evident in Officer Barton's sketch of the object. Not only did Barton's sketch detail the locations of several lights, but also noted the location of what can only be described as an array of lights at the rear of the craft. Barton's diagram also makes fairly precise estimates, as to the craft's length and width, something that a civilian witness might certainly do, although perhaps not to that level of detail.

Officers Craig Stevens and Ed Barton;
Witnesses Melvern Noll and Steven Wonnacott

*Courtesy Black Vault Encyclopedia Project
via Riverfront Times, St. Louis, Missouri*

Millstadt, Illinois as a high credibility case

The Millstadt, Illinois event qualifies, in the opinion of this writer, as a high credibility case for several reasons, most notably, though not exclusively, due to the fact that it was not only a multiple witness event, but that, of those multiple witnesses, several were law enforcement officers.

In addition to multiple members of law enforcement reporting the same (or a similar) object, it is important to note that these individuals made their sightings independent of one another, at different times and in fact, different communities altogether. In the most basic of terms, the odds of multiple witnesses, over a series of miles, all either misidentifying the same object or confabulating similar accounts strike this writer as being long indeed. Though unconventional, would it not seem more reasonable that multiple

members of various police agencies did, in fact, see an anomalous aircraft on the early morning of January 5, 2000?

It is important to note that, while several officers, either on-scene or by dispatch, made references to UFOs or spacecraft, this was, to the best of my knowledge, simply as an analogy for what the officer was seeing – a means by which to provide a frame of reference.

This, in itself, is not entirely unusual. Witnesses often make such analogies when recounting their experiences.

Lastly, the Millstadt event is highly credible, in the mind of this writer, due to the number of secondary sources of information potentially available.

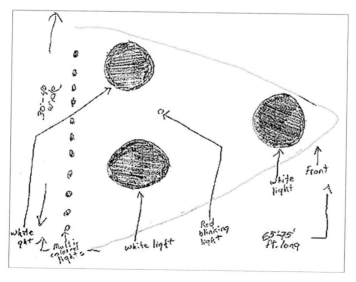

Sketch of object by Officer Ed Barton;
Lebanon, Illinois Police Department

Courtesy Black Vault Encyclopedia Project

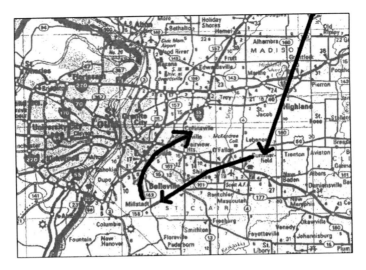

Map showing the area of the event, including Scott Air Force Base

Courtesy David Marler

Not only are the police dispatch transcripts and recordings available, but the case is also bolstered by the fact that the general area in which the event occurred includes not only Scott Air Force Base, which was queried by David Marler, then State Director for Illinois MUFON, but also Lambert International Airport in St. Louis, Missouri, which was also in the flight path of the object on the evening in question.

It is certainly possible that the object in question was an as-yet-unknown advanced military craft. The close proximity to Scott Air Force Base and flight path of the object make this a distinct possibility, especially in the minds of the police witnesses that night.

However, in his March, 2000 article for the *MUFON UFO Journal*, Illinois State Director Marler addresses the issues surrounding the Scott Air Force Base connection by saying:

> "*I find it hard to imagine that a large unknown object could fly under 1000 feet within two miles of the base without causing some concerns. I believe they may have*

more detailed information concerning this event. It is a distinct possibility that the object witnessed was a military craft of some type. Unfortunately, we do not have any evidence to support or refute this idea."

As noted by Marler, a number of details of the Millstadt event simply do not seem to add up.

Firstly, an aircraft of the size reported, flying at an altitude of less than one thousand feet would almost assuredly have generated a significant amount of noise. In contrast, the craft in question was reported to have emitted little or no sound.

Secondly, one can only assume that, regardless of origin, the craft in question would almost certainly have been tracked by personnel at Scott Air Force Base, less than two miles from the sighting location.

According to Marler's original case report, as published in the March 2000 *MUFON UFO Journal,*

> *"Due to the location of Scott Air Force Base in relation to the UFO sightings, I found it necessary to contact them. Although I was not optimistic, I felt I should pursue every possible source of information. I wrote a letter of introduction, requesting answers to four important questions concerning the recent UFO sighting. ...I promptly received a courteous, albeit uninformative, letter of response indicating that (1) the only calls received at the base were from the media, (2) no ground observers at the base have come forward with reports, (3) the base did not track any objects on radar, since radar services are provided by the FAA at Lambert-St. Louis International Airport, and (4) the base was not operating any craft resembling the reports".*

While this writer has no reason to doubt Marler's report, the response from Scott Air Force Base does seem somewhat curious. While I do not doubt that the base received calls from the media, that ground observers may not have come forward, or that radar services are provided by nearby Lambert International Airport in St. Louis, it is certainly curious that Scott's letter to Marler makes no mention of any contact from police dispatch, as records from the St. Clair County, Illinois Sheriff's Department clearly show officers from Shiloh, Illinois requested that dispatch contact Scott Air Force Base.

The Millstadt event as a 'special circumstance' case

The Millstadt, Illinois event can be seen, in the opinion of this writer, as being a "special circumstance" case due to the fact that it seemingly tends to corroborate previous events in the history of the phenomenon, separated by time and distance, as well as incorporating multiple witnesses over a wide area, as well as being documented in copious detail by one of the many witnesses.

It is important to mention, specifically with regard to the many police witnesses in this incident, that each of the witnesses involved – chiefly Officer Ed Barton of Lebanon and Officer Craig Stevens of Dupo – reported the incident independent of one another and, when interviewed by researchers, each produced strikingly similar accounts of the object in striking detail.

The Millstadt event did seem, in this writer's opinion, to validate the numerous sighting reports of black, delta-shaped craft which have been common in this field for more than two decades. While it is true that these types of reports were not uncommon, the Millstadt event presented a set of witnesses whose credibility was nearly unimpeachable, being law enforcement professionals.

The discrepancies between witness reports is noticeable, though not uncommon, even in cases involving professional witnesses such as law enforcement officers.

The Millstadt event can also be seen as a unique example, in my view, due to the secondary evidence produced on the morning of January 5, by way of the police dispatch recordings and transcripts.

As mentioned previously, these transcripts are particularly valuable when reconstructing the series of events between approximately 4:10 AM and shortly after 4:30, as the two-way dispatch recordings were able to provide researchers such as MUFON's David Marler with an up-to-the minute accounting of events.

March, 2000 MUFON UFO Journal, which featured the Millstadt event
Courtesy Mutual UFO Network

The Millstadt, Illinois event in historical context

The Millstadt, Illinois event, in my view, should be viewed as being particularly significant when viewed in historical context for several reasons outlined previously.

The Millstadt event offers a mix of civilian, law enforcement, and, potentially, military witnesses to an event that encompassed several jurisdictions and spanned two states. For this reason, this writer is inclined to liken the case to the well-known and much publicized

ıf Washington National Airport and the surrounding area
ᵤ ᴄne summer of 1952.

Though not often considered to be a photographic evidence case, the Millstadt event does possess a photographic component, although it is of poor quality due to extreme cold temperatures on the morning in question, which led to a camera malfunction, which, in turn, is responsible for the generally poor condition of the print.

Even if the photograph in question is discounted entirely, this case is still a strong one, given the secondary reports made by two of the police officers involved, as well as the recordings made by police dispatch, which, in essence, serve to document the event and provide a relatively accurate timeline of events.

Taking these details into account, it is the opinion of this writer that the Millstadt, Illinois event of January 5, 2000 will stand as a highly credible case for years to come even though, as of this writing, the best and most complete determination that can be made regarding this case may well be "inconclusive".

18. The Chicago OHare Airport Event

On November 7, 2006, United Airlines employees observed a disk-shaped object hovering over O'Hare International Airport, in Chicago, Illinois. It then shot off at a high rate of speed, leaving a gaping hole in the overcast cloud, revealing the blue sky above.

The United Airlines employees called their manager to advise that on object was hovering over gate C17. The manager also saw the object and proceeded to contact the Federal Aviation Administration (FAA). The ground crew alerted the pilots onboard a flight being pushed back from the ramp. They too confirmed the sighting. All reported the object as disk-shaped and resembled "a rotating Frisbee". It is noteworthy to mention that this same effect was described by the witnesses in the Vancouver Island UFO case back in the early 1980s.

This story gained worldwide attention from the media, from the internet, to radio, television and cable. In fact, the *Chicago Tribune* reported that their website had more hits on January 1, 2007 (over one million) than any other day in their online history.

The first report was received by the National UFO Reporting Center (NUFORC). Its director, Peter Davenport, asked local Chicago investigative reporter Jon Hilkevitch to investigate the story. Any object which flies in restricted airspace at any international airport without radio contact, transponders, or permission to be in that airspace, could potentially be a safety hazard.

A UAL pilot, who taxies aircraft around O'Hare, also reported the object, which he described as *"a clearly discernable gray disk with atmospheric distortion at the bottom, like heat rising off a desert road*". He also mentioned that where the object had been, there was a perfect circle in the cloud layer. This pilot was interviewed on television, including CNN. What is interesting here is that the pilot was not pressured to remain silent. Historically, pilots who reported UFOs were considered to have committed "career suicide". Perhaps the view of ufology by the mainstream media truly is changing.

Photographic and video evidence of the O' Hare incident

Photographic and video evidence stemming from the O'Hare International Airport event has been in short supply to say the very least, but in the recent past, questionable photographs, purportedly from the O'Hare event, have entered the public domain, by way of the popular website, *UFOCasebook.com*.

While the photographs are questionable at best, however, given compression of the image and the low image quality, it seems to me that the photographs (or video screen captures, whichever the case might be) are generally consistent with having been captured using the camera function common to many cellular phones.

Photographs which purportedly show the Chicago O'Hare UFO.
Courtesy BJ Booth / UFO Casebook

To my knowledge, there has been no analysis done on any of the photographs or video material purportedly shot during the O'Hare event, primarily as the site of the incident was one of the busiest international airports not only in the United States, but in the world, and as such, the odds of locating the individuals who flew either into or out of O'Hare International Airport on November 7, 2006 are quite long indeed.

If, however, a witness could be definitively located who was willing to submit materials in their possession to scientific testing, I believe a thorough analysis of any available photo or video evidence could be quite enlightening, due to the security concerns highlighted by this particular event.

As mentioned previously, only a few select photographs of dubious origin have surfaced to date. Perhaps, in time, better information will become readily available that might allow investigators of all backgrounds to take a further look into the O'Hare event.

The O'Hare International Airport sighting as a high credibility case

The O' Hare International Airport event, in my view, should be held as a high credibility case, alongside others in the history of ufology, such as the Washington National Airport cases of 1952, given the fact that, as with the Washington event, the object in the O'Hare case seemingly entered the restricted airspace of the airport, without raising any sort of alarm by way of radar.

The O'Hare event should also be seen as highly credible given the media scrutiny under which the event occurred. Given the sheer number of potential eyewitnesses to the case, the O'Hare UFO event could very well be, providing that witnesses be located or otherwise come forward, one of the single-best documented cases in modern history.

O'Hare International Airport, Chicago
Courtesy Carter Dayne / IStockPhoto

The Federal Aviation Administration and news media as sources of evidence

As is typical, the FAA denied they received any reports on the O'Hare incident but changed their story on learning an FOIA request had been filed. Their official said:

"The FAA is not conducting a further investigation. Our theory on this is that it was a weather phenomenon. That night was a perfect atmospheric condition in terms of low ceiling and a lot of airport lights. When the lights shine up into the clouds, sometimes you can see funny things. That's our take on it."

This, as an official statement from the FAA, is both expected and somewhat surprising.

Firstly, it is expected and predictable in the respect that the FAA's statement following the O'Hare event fits a formulaic pattern, which seems to reach as far back as the Roswell event: Issue a statement, but issue a statement that allows you as the official spokesperson to divest yourself of the situation as expeditiously as possible.

Secondly, the FAA statement is quite surprising at the very same time. It is, at least to the mind of this writer, highly curious that a purported UFO event could occur in the third most populous city in the United States, at one of the nation's most prominent and busiest hubs of air travel and have the Federal Aviation Administration choose to dismiss the event out of hand, seemingly without so much as a cursory investigation, especially when one may be required, given that an unidentified craft did breach the airspace over O'Hare. Moreover, how is the Federal Aviation Administration, without benefit of even the most cursory of investigations, capable of determining the cause of the event as being natural phenomena?

As Illinois MUFON State Director Sam Maranto queries in the February 2007 issue of the *MUFON UFO Journal:*

"This raises the following questions: how does weather account for a solid object of distinguishable color and shape exhibiting erratic hovering motion which can then bolt through the cloud cover leaving a cookie cutter type hole? How does it account for an object seen from multiple vantage points by multiple witnesses who by virtue of their aviation jobs know what should be in that airspace and

what should not be there? The answer to these questions is that weather conditions cannot account for this sighting and the weather explanation is just complete and utter nonsense."

During an interview with Peter Davenport, Director of the National UFO Reporting Center, and NBC Consultant and noted UFO debunker James Oberg, MSNBC Science Reporter Alan Boyle suggested that the UFO was a lenticular cloud.

For those unfamiliar with cloud types, a lenticular cloud is a stationary, lens-shaped cloud that forms at high altitudes normally perpendicular to the wind direction[44]. Lenticular clouds have been mistaken for UFOs because they have a lens (or disk) shape and smooth saucer-like appearance. However, these cloud types generally form on the lee side of mountains and normally in the western United States, rarely as far inland as the Chicagoland area.

A lenticular cloud, even on the rare occasion it has been seen in Chicago, could not bolt through a cloud layer and create the perfect circular hole that witnesses reported accompanied the O'Hare event.

Lenticular cloud formation
Courtesy National Oceanic and Atmospheric Administration

James Oberg also made a bizarre statement about the reliability of airline pilots as witnesses to aviation events. His statement was that the *"NTSB (National Transportation Safety Board) investigators say that the worst observers of an aviation accident are aviation personnel"*. He added that pilots want to understand what happened and in their initial perceptions and retelling of the events, will usually stress the facts supporting their interpretation. Mr. Oberg is certainly entitled to his opinion; however, I feel that Mr. Oberg's statements might be colored, for want of a better term, by his desire to reach a particular conclusion. There is no doubt that Oberg is a bright individual, exceptionally well-versed in his particular field. This is all well and good; however his many years as an ardent UFO debunker make Mr. Oberg's statements highly questionable in the opinion of this writer. A skeptic, by definition, has kept an open mind to the evidence. Individuals such as Mr. Oberg, however, have seemingly come to an intractable conclusion, and no amount of evidence or well-reasoned argument can sway them from their position.

In contrast, it is important to again revisit the words of Illinois MUFON State Director Sam Maranto who was directly involved with the research into the O'Hare event at the time:

> *"If pilots are not able to make valid, real-time observations of shapes, color and motion in the sky, they cannot be trusted with the safety of airline passengers. Real time observations and analyses are constant critical functions, routinely performed by pilots. No one does this better – period!"*

If pilots, whom many consider to be among the most credible and reliable of witnesses, are, as Oberg contends, such poor and unreliable witnesses, one would expect to see air travel figures drop off significantly with each passing year, however, this has not been the case. While it is true that air travel figures have seen a drop in recent years, the cause of this decline is seen by many as being economic in nature, and not the result of an air accident or negligence on the part of commercial pilots, by any stretch of the imagination.

The O'Hare Sighting and its aftermath

The O'Hare International Airport UFO event had a profound effect on the field of UFO research and, perhaps more to the point, the treatment of the subject in the mainstream press. The O'Hare sighting and subsequent media attention by way of the *Chicago Tribune* and other media outlets undoubtedly influenced the media coverage of future UFO events, such as the January, 2008 events at Stephenville, Texas.

Put very simply, the web traffic driven to the *Tribune* website following the O'Hare event – in excess of one million unique hits – showed the rest of the journalistic world, whether the medium was television, print, or web-based, that stories pertaining to UFOs were of serious interest to members of the general public. In journalism, the formula for success is simple. Cover the stories of interest to your readers. In the very simplest of terms, the O'Hare UFO sighting had made UFOs not only relevant as a topic of discussion amongst the general public throughout the world, but UFOs had become big business to those in the media.

The O'Hare International Airport event in historical context

The O'Hare International Airport case has the potential, in the opinion of this writer, to occupy a significant place in the history of ufology in the twenty-first century specifically because of the massive amount of media attention generated by the case. There is no debating the fact that the O'Hare event, whatever it may have been, did generate interest significant enough to virtually crash the website of the *Chicago Tribune* on January 1, 2007.

In my estimation, the O'Hare case is much less a photographic evidence case (while that component is certainly present) and is much more a commentary on the nature of the media.

Moving forward, I believe researchers will view the O'Hare event as one of several cases occurring in the early part of the twenty-first century which helped cement the nature of the UFO phenomenon in the minds of the general public as being much more a matter of hard science than science fiction.

While the O'Hare event remains largely speculative at present, due, primarily, to the lack of firsthand witnesses willing to go on record, it seems that there is potentially a wealth of evidence available, as mentioned previously, due to O'Hare International Airport serving tens of millions of passengers annually.

Though civilian researchers may not yet have a definitive answer for those who wonder what was seen over the C concourse of the airport on November 7, 2006, I believe all parties concerned can come to agreement on the fact that it was certainly something that does not fit the blanket prosaic explanation offered up by the FAA, and as such, deserves further study in the years ahead.

(Endnotes)

[44] http://en.wikipedia.org/wiki/Lenticular_Clouds

19. In Summation

The burden of proof, in non-criminal court proceedings, is not *"beyond a reasonable doubt"*, as is the common misconception, but is, instead, *"by a preponderance of the evidence"*. By legal definition, this standard is met if what is proposed has a higher probability of occurring than not.

Based on the cases covered in the preceding chapters, as well as the documentary evidence associated with said cases, I believe that the evidence is overwhelmingly in support of an *"unknown"* determination. The cases profiled herein have each individually withstood scientific investigation and scrutiny time and again, some for longer than six decades.

The Invisible College

"The Invisible College" was the name given by astronomer and physicist Dr. J. Allen Hynek to the underground community of UFO researchers within the academic community as early as the mid-1960s. In the coining of this term, it was Hynek's contention that enough scholarly, academic interest existed in what he called *"the UFO problem"* to staff a major university. This phraseology is particularly telling as Dr. Hynek viewed UFOs through the lens of science. In so doing, he viewed *"the UFO problem"* as any other in science –a puzzle to be solved; a set of circumstances to which there was an analytical and deductive solution.

Hynek himself is a prime example of the persuasive nature of the evidence available on the UFO subject. Initially a scientific consultant to the United States Air Force's Project Blue Book, the military's official investigation into the UFO subject, Hynek became a staunch supporter of UFO research after his personal experiences while investigating a series of UFO reports over the State of Michigan in 1966. Hynek would take an active role in UFO research and maintain that role, authoring scholarly papers and books, and speaking worldwide until his death in 1986, at age 76.

The Michigan UFO events were a watershed in Hynek's career as a consultant to the US military. Faced with a mounting number of case reports that defied conventional explanation, Hynek became firmly convinced that there was enough evidence to warrant continued scientific study.

Following the Michigan events, Dr. Hynek stated publicly that it was in the interests of science to objectively and scientifically study the UFO phenomenon, rather than continue the active campaign of debunking the subject, which Hynek said had been prevalent in the Air Force's investigation since its inception in the late 1940s.

Dr. Hynek, of course, was not the only academic involved in ufology, simply one of the first to advocate this position. Since Hynek's time, many career academics, including tenured professors at major American universities, have publicly taken the position that the research being done into the UFO subject adheres to accepted scientific practices, and should be studied accordingly.

One of the most notable academics in modern ufology is Dr. David Jacobs, associate professor of History at Temple University in Philadelphia. Dr. Jacobs's stature is significant in that his doctoral dissertation, written in 1973 while at the University of Wisconsin-Madison, concerned the controversy surrounding the UFO subject in America. A revision to this dissertation was published in 1975 as *The UFO Controversy in America* by Indiana University Press. To this day, this remains the sole sympathetic book published on the UFO topic by an academic press.

The academic involvement in the UFO field is as varied as the individuals involved. The Mutual UFO Network's Board of

Consultants, those who hold a Master of Sciences/Arts or doctorate degree, includes some sixty-plus different scientific and technical disciplines.

Additionally, there are a small number of American institutions, including Temple University, which offer, or have offered in the past, academic courses dealing with the UFO subject. While the number of schools currently offering these courses as an official part of their curriculum is anemic at best, one can only presume that this number will increase in coming years, as the current paradigm continues to shift and research into unexplained phenomena, such as UFOs, becomes an ever larger part of the global reality.

It is interesting to note that a number of developed or developing nations, as previously mentioned, treat research into the UFO topic as a simple matter of science, rather than with the skepticism presently seen in the United States. Research into *"the UFO problem"*, as Hynek termed it, is ongoing worldwide, receiving serious governmental attention and, in some cases, government funding, in countries such as Canada, Mexico, France, the United Kingdom, and China.

As recently as 2007, the government of France declassified thousands of pages of research into the UFO subject, in addition to a previously released 1999 report titled, in English, *UFOs and Defense: What must We be Prepared for?* The 90 page report, issued by the French Institute of Higher Studies for National Defense, was sent to a number of high-ranking French officials, including President Jacques Chirac and Prime Minister Lionel Jospin.

While closed minded individuals are still actively debunking the UFO phenomenon to this day, these individuals are now in the minority. According to a September, 2002 public opinion poll conducted by the Roper organization, 56% of those surveyed said they believed UFOs to be a very real phenomenon. These types of numbers suggest that a majority of the general public acknowledges both the possibility and probability of the existence of a technology beyond current understanding.

This same public opinion poll shows that a majority of the general public believes that information pertaining to and/or evidence of the reality of UFOs is being suppressed from circulation in the mainstream media. When asked this question, 72% of respondents said that they believed that the government was not sharing all available information on the subject with its citizens. 60%, or approximately 600 of the over 1000 people surveyed, said that they believed whatever information the United States government may have on the UFO subject should be released into the public domain, provided that this information did not compromise United States national security interests.

With such striking numbers of private citizens coming to a consensus on this issue, it appears that the gap between the United States and other industrial nations around the globe, with respect to UFO research, may be narrowing slightly.

Photographic and video evidence and its forensic value

From the McMinnville, Oregon photos taken in 1950 to the sightings over O'Hare International Airport in 2006, photographic and video evidence cases, while in the minority of reported cases as a whole, are vital pieces of the proverbial puzzle.

Cases such as McMinnville and the Rex Heflin photos, taken in Santa Ana, California in 1965, serve as vital controls when evaluating modern-day photographs.

Though the objects seen in these photographs might be unknown or as yet unidentified, the additional data available on these photographs allow them to be used, in a manner of speaking, as forensic controls in the study and analysis of UFO photographs of a more recent vintage. As with any scientific discipline, there is a direct correlation between present research and that which has come before. Photographs from the early days of ufology, the 1940s through late 1960s, can be used as a basis for comparison with modern photographs, comparing similar features or, in the case of misidentification, contrasting an as yet unidentified feature with a misidentified photo, which is a known terrestrial object.

While these techniques may not definitively identify all questioned UFO photographs, it does firmly root the field of ufology in the bedrock of accepted science. This is invaluable because the scientific method dictates that in order to be valid, experiments must be reproducible under controlled conditions.

The 1991 Space Shuttle Discovery UFO video footage is a prime example of accepted science and ufology coexisting side by side. The scientific controls in this case were established due to the fact that the video in question came directly from a satellite feed from NASA.

Longtime UFO researcher Dr. Jack Kasher, himself a university physics professor, was able to apply a scientific methodology to the questioned object seen in this video and determine that it was moving both at speeds that a human pilot could not withstand, as well as possessing physical properties not shared by conventional air or spacecraft.

Scholarly investigative reports, such as Kasher's scientific analysis of the videotape of the Space Shuttle Discovery (flight STS-48), are the byproducts which the scientific community expects of serious research. In this particular case, the analysis of the STS-48 event, published in 1995 through a grant from the non-profit Fund for UFO Research (FUFOR), covered some 105 pages, including all the requisite calculations and reference material needed for Kasher's research to be appropriately vetted by the scientific community.

While some in the academic community do not see *"the UFO problem"* as worthy of scientific inquiry, tens of thousands of cases, consisting of hundreds of thousands of pages, as well as other documentary and anecdotal evidence, clearly meet the burden of proof.

Unexplained phenomena research and researchers in historical context

The field of ufology, a field that is still emerging after nearly three quarters of a century, has already produced a significant number of researchers who, by their body of work, have come to embody this field of study.

Just as with any field, the research done by some of ufology's pioneers, men like Allen Hynek, Donald Keyhoe, James McDonald, and John Schuessler, among others, lays the groundwork for those who will carry the proverbial torch in the next generation.

As some of the early research in this field dates from the 1940s, many of the first generation researchers are no longer living. In these cases, anecdotal and hearsay evidence is, many times, the best, if not only available evidence.

From a scientific point of view, first-hand documentation and testimony is always the best evidence. When, as noted, the initial researchers have since passed away, their personal research becomes their voice, allowing generations too young to have known them personally the opportunity to still benefit from their discoveries.

One example of a modern ufologist's research being preserved for posterity is the case of Dr. James McDonald. A tenured professor of physics at the University of Arizona at Tucson, Dr. McDonald was involved in several early UFO cases, including the investigation of the 1965 Santa Ana, California photographs shot by California Highway Department engineer Rex Heflin.

Following Dr. McDonald's death in 1971, his personal papers, published articles and other research stayed in the possession of the McDonald family until a grant from the Fund for UFO Research allowed for their archival preservation. McDonald's personal archive, totaling some 60,000 printed pages, as well as other materials such as reel to reel tape recordings, were subsequently donated to the Special Collections Section at the University of

Arizona Library in Tucson, Arizona, where McDonald spent the bulk of his professional career.

By following the lead of the Fund for UFO Research and preserving other private document collections, future researchers are guaranteed access to materials which, from an academic standpoint, are priceless.

Unexplained phenomena research and the mainstream scientific community

As mentioned previously, the field of ufology offers an alternative perspective to the sciences, rather than, as some claim, posing a threat to scientific inquiry. UFO researchers adhere to scientific protocols in their research, employ the scientific method, and adhere to the methodologies commonly found in other fields of research. These processes are all necessary to ensure that research into unexplained phenomena is taken as seriously as is possible by the mainstream scientific community.

There is a symbiotic relationship of sorts between the field of UFO research and the mainstream scientific community as a whole. Gradual advances in science and technology allow those of us involved in UFO field research to do our work with an even greater degree of accuracy than ever before, while the scientific community, or, more correctly, a percentage thereof, needs the UFO research community to continue the ongoing debate on the subject, which, in turn, it can be argued, leads to an increased interest in space exploration, as well as advances in the aerospace industry.

The next step: research, public relations, and the implications to mainstream science

Finally, as research into Unconventional Flying Objects enters the twenty-first century, questions can be posed as to what the logical extensions of research into the subject are, along with the implications to mainstream science.

At present, the data collected by a research organization such as the Mutual UFO Network is, primarily, proprietary in nature. While general research information is available to the public, a majority of cases, some dating back to the late 1960s, remain in storage, their details rarely seen by the general public. This is due to the fact that, while some information, such as date and location of a particular event, is a matter of public record, a number of files remain confidential, chiefly due to witness anonymity requests.

Promoting this research and disseminating it so as to capture public attention has rarely been done to any great extent in years past. To many, this field is the subject only of the occasional cable television special or Internet news story.

When tastefully done, news coverage can spur somewhat skeptical individuals to report their previous experiences.

As mentioned previously, an organization such as the Mutual UFO Network, benefiting from the expertise of research specialists and consultants representing over 100 different technical disciplines, projects the very image of science. These are the individuals who, in many cases, find themselves fielding the technical and scientific questions related to ongoing UFO research.

As recently as January 2008, large-scale UFO sightings involving hundreds of witnesses captured major media attention, including prime time news coverage by mainstream news media such as Fox News and CNN. While not unheard of in the UFO field, this type of mainstream media exposure is certainly uncommon.

In recent memory, only the Phoenix Lights case in 1997 drew the mainstream media coverage of the January, 2008 Stephenville, Texas sightings. The interest in the Texas sightings was so substantial that the Mutual UFO Network received as many sighting reports on this one incident as it usually receives in a given month for the entire United States.

In a recent news story surrounding that case, the Dallas-Fort Worth CBS television affiliate stated that, according to a survey conducted in January 2008, some 14% of respondents said that they had had some form of UFO sighting or encounter. The US Census Bureau, as of July 2007, puts the population of the United States at approximately 301.1 million people. If this 14% figure is, in fact, representative of the entire US population, this would put the number of UFO experiencers presently living in the United States at approximately 42.1 million people.

Positive media relations, media appearances, and public statements pertaining to the UFO subject is perhaps the most fundamental responsibility of any serious ufologist today. By representing the field of ufology in a professional manner, the positive public perception of the entire research field increases exponentially.

Lastly, the success of the field of ufology has far-reaching implications to the mainstream scientific establishment. While there are a certain number of empirical scientists for whom no amount of anecdotal data or supporting research will ever be enough, there are several high profile individuals within the scientific community, specifically physicists such as Dr. Michio Kaku of New York University and Dr. Stephen Hawking of the University of Cambridge, England, who have spoken at length on the UFO problem and its value to the academic scientific community.

Dr. Kaku, himself a theoretical physicist, has openly advocated for the study of UFO sightings by the mainstream scientific community for several years. Discussing the need for scientific investigation into this subject, Dr. Kaku stated:

> *"You simply cannot dismiss the possibility that some of these UFO sightings are actually sightings from some object created by advanced civilization, a civilization far out in space, a civilization perhaps millions of years ahead of us in technology. You simply cannot discount that possibility... when you look at the handful, the handful of cases that cannot easily be dismissed, this is worthy*

of scientific investigation. Maybe there's nothing there. However, on the off chance that there is something there that could literally change the course of human history, so I say, let the investigation began."

It is pure supposition, at this point in time, to think that the possibility exists to recover technology from a UFO event. That being said, trained scientists like Dr. Stephen Hawking and Dr. Michio Kaku have suggested, like Dr J. Allen Hynek before them, that the enigmatic field of UFO research, with or without accompanying documentary evidence, poses such a unique opportunity for scientific discovery that it warrants immediate and rigorous study by the scientific establishment.

After six decades of study and scientific advancement, the field of UFO research, specifically where photographic images are involved, has raised as many questions as it has answered. Perhaps we are simply lacking the final piece of the puzzle, which may bring the entire phenomenon into sharp focus.

About the Author

Nicholas D. Roesler has been involved with paranormal research, specifically the study of UFOs, since 1998. In that time, he has worked extensively with the Mutual UFO Network (MUFON) and Smoking Gun Research Agency (SGRA). He has previously served as both Assistant and State Director for Wisconsin MUFON and presently is the Staff Photographer for the *MUFON UFO Journal.*

Mr. Roesler has frequently lectured on photographic evidence UFO cases across the United States.

In his professional life, Mr. Roesler has been employed in the criminal justice field since 1999.

He presently resides in Milwaukee, Wisconsin.

Breinigsville, PA USA
18 October 2009

226012BV00003B/4/P